Neighbourhood Planning

This book carries out an in-depth investigation of a neighborhood planning process that engages critically with the issues surrounding articulation of local concerns in a strategic manner and the prospects of implementing 'bottom up' community initiatives successfully.

It highlights the dynamics involved in shaping the content of a neighbourhood plan and the implications of the different ways in which a place is constructed. The book challenges the notions of a singular place that is described in a neighbourhood plan. It examines conceptual, thematic, strategic and performative constructions of place and the capacity for neighbourhood plans to be developed within this context. It explores the value of connecting the formulation of a neighbourhood plan with the emergence of a relevant local plan, allowing for more meaningful local influence on strategic policymaking.

With first-hand insights on neighbourhood planning, this book offers a novel contribution to the fields of planning, urban studies, and urban geography.

Janet Banfield teaches human geography at Oxford University. Janet's previous career included environmental consultancy in the private sector and corporate policy in local government. This book brings these professional interests together, reflecting upon her time and experience as Vice-Chair for her local neighbourhood plan.

Routledge Studies in Urbanism and the City

Citizenship and Infrastructure
Practices and Identities of Citizens and the State
Edited by Charlotte Lemanski

Pedagogies of Urban Mobilities
Kim Kullman

Balkanization and Global Politics
Remaking Cities and Architecture
Nikolina Bobic

Cities and Dialogue
The Public Life of Knowledge
Jamie O'Brien

The Walkable City
Jennie Middleton

Urban Restructuring, Power and Capitalism in the Tourist City
Contested Terrains of Marrakesh
Khalid Madhi

Ethnic Spatial Segregation in European Cities
Hans Skifter Andersen

Big Data, Code and the Discrete City
Shaping Public Realms
Silvio Carta

Neighbourhood Planning
Place, Space and Politics
Janet Banfield

For more information about this series, please visit www.routledge.com/Routledge-Studies-in-Urbanism-and-the-City/book-series/RSUC

Neighbourhood Planning

Place, Space and Politics

Janet Banfield

LONDON AND NEW YORK

First published 2020
by Routledge
2 Park Square, Milton Park, Abingdon, Oxon OX14 4RN

and by Routledge
605 Third Avenue, New York, NY 10017

First issued in paperback 2021

Routledge is an imprint of the Taylor & Francis Group, an informa business

British Library Cataloguing-in-Publication Data
A catalogue record for this book is available from the British Library

Library of Congress Cataloging-in-Publication Data
A catalog record for this book has been requested

ISBN 13: 978-0-367-77754-8 (pbk)
ISBN 13: 978-0-367-19994-4 (hbk)

Typeset in Times New Roman
by Apex CoVantage, LLC

For neighbourhood planners everywhere: I hope it's worth it!

Contents

List of figures ix
List of tables x
Acknowledgements xi
List of abbreviations xiii

Introduction 1
Introduction 1
Context, aims and audience 2
Analytical themes and chapter outline 5
Conclusion 9

1 Localism and neighbourhood planning 13
Introduction 13
Localism 13
Neighbourhood plan context 19
Wootton and St Helen Without Neighbourhood Plan 23
Neighbourhood plan and local plan 26
Conclusion 28

2 Conceptual constructions of place 31
Introduction 31
Dimensional and networked understandings of space 31
Performative and affective understandings of place 38
Conclusion 44

3 Thematic constructions of place 48
Introduction 48
The National Planning Policy Framework (NPPF) 49
Environment 52
Green Belt 56
Sustainability 60
Conclusion 63

4 Strategic constructions of place 68
Introduction 68
Strategy as scalar: questioning the strategic 69
Strategy as spatial: site or area? 73
Strategy as temporal: place or process? 78
Conclusion 82

5 Performative constructions of place 86
Introduction 86
Political constructions of place 87
Practical constructions of place 91
Presentational constructions of place 95
Conclusion 100

6 Reflections and projections 103
Introduction 103
Theorising place 104
Legitimacy 110
Prospects 115
Conclusion 121

Conclusion 125
Introduction 125
WSHWNP update 129
Thematic insights and proposals 133
Conclusion 139

Index 142

Figures

1.1	Locating the WSHW Neighbourhood Plan	21
2.1	Spatial imaginaries of neighbourhood planning	37
6.1	Legitimacy framework for neighbourhood planning	112
6.2	Chimerical instability delimiting a space of negotiation	119

Tables

3.1 NPPF Sustainable Development Aims 2012 and 2018 50
3.2 Comparison of understandings of the environment 54
3.3 Comparison of understandings of the Green Belt 57
3.4 Comparison of understandings of sustainability 61
6.1 Theorising the spatial experiences of neighbourhood planning 107

Acknowledgements

Without the tireless efforts of everyone involved with the Wootton and St Helen Without Neighbourhood Plan, I would have no content for this book, so my sincere gratitude goes to all those on the steering group for sticking with it and for putting up with me for the past two years; and to members of both Wootton Parish Council and St Helen Without Parish Council for their enduring and strong support for both the neighbourhood plan and for me personally. I am also grateful to the volunteers and local businesses who have helped to keep the neighbourhood planning process on track and the community informed and engaged, and to everybody who has fed into the development of the neighbourhood plan, whether through consultation responses, provision of technical advice and services, or proof-reading, designing and preparing both the plan and its associated website. I hope that the narrative in these pages does justice to all the hard work, emotional energy, community commitment and expertise that has been invested in our neighbourhood plan.

Thank you also, to Faye Leerink, Ruth Anderson and Nonita Saha at Routledge, for their guidance and support in bringing this volume to publication, and to the anonymous reviewers who provided such supportive, constructive and encouraging advice on how to strengthen the text and optimise its potential contribution, both academically and practically.

As ever, I am so appreciative of the support of Mum, Dad, Spoon and countless others for providing such resilient sounding boards for my stress and frustration about the neighbourhood plan and for providing keen enthusiasm for turning this aggravation into a positive publication opportunity. Specific mentions go to Mark Burgess, Rob Dunford, Jamie Lorimer, Fiona McConnell, Alex Money and to Carole and Mick Page for their patient tolerance of my grumbles about the neighbourhood planning process, for their encouragement both to persist with the plan and to pursue this publication opportunity, and for the many laughs we have had along the way, which have been such a valuable counterfoil for all the inconvenience, disappointment and anger involved in neighbourhood planning. Thanks also go to all the wonderful, friendly and community-minded people I have had the pleasure to meet and get to know through my involvement with the neighbourhood plan, without whose company and camaraderie the whole process would have been utterly miserable. On this note, another special mention goes to Freddie, whose entertaining antics have lightened many a mood and many a meeting.

Finally, thanks must go to the Vale of White Horse District Council. I might disagree vehemently with the local authority's position and some of its actions with respect to our neighbourhood plan, but I do not question the proficiency or professionalism of any of the individuals involved. Consequently, I am grateful to the 'Vale' for their sustained engagement with the neighbourhood plan, which on the one hand gave me so much to write about and on the other hand lit a fire in my belly to persevere in writing it.

Abbreviations

DCLG	Department for Communities and Local Government
DEFRA	Department for the Environment, Food and Rural Affairs
DIO	Defence Infrastructure Organisation
HDA	Hankinson Duckett Associates
HRA	Habitats Regulations Assessment
LPA	Local Planning Authority
LPP1	Local Plan Part 1
LPP2	Local Plan Part 2
MHCLG	Ministry of Housing, Communities and Local Government
MoD	Ministry of Defence
NPIERS	Neighbourhood Planning Independent Examiner Referral Service
NPPF	National Planning Policy Framework
NPSG	Neighbourhood Plan Steering Group
SEA	Strategic Environmental Assessment
TCPA	Town and Country Planning Association
TDRC	Thomas Design Regeneration and Consultation Ltd
VWHDC	Vale of White Horse District Council
WSHW	Wootton and St Helen Without
WSHWNP	Wootton and St Helen Without Neighbourhood Plan
WSHWNPSG	Wootton and St Helen Without Neighbourhood Plan Steering Group

Introduction

Introduction

2016 was an auspicious year for neighbourhood planning in England. In October that year, for the first time, a neighbourhood plan was rejected at referendum. This was a notable event itself, but it is even more remarkable that the steering group responsible for the development of that neighbourhood plan had actively campaigned for its rejection (Milne, 2016). This bizarre turn of events apparently sprang from the perception that changes made to the plan by the local planning authority (LPA) following examination without the involvement of the steering group left the neighbourhood plan no longer representing the aspirations of the local community. In other words, it was no longer their plan and it no longer served their purposes.

This incident speaks to the very essence of neighbourhood planning and demonstrates the political wrangling between communities and local authorities that frequently features in academic literature on neighbourhood planning. At its heart, neighbourhood planning is an explicitly geographical and community-grounded activity that enables residents to influence development by participating in the planning process at a very local level. It is an exercise in community place-making but it takes place in the context of a hierarchical planning system that simultaneously retains power at the level of the LPA while also seeking to transfer power to local people.

Neighbourhood planning is a new tier in the planning system which was introduced under the Localism Act 2011 to give power to local people within a broader shift towards the devolution of power from central government to local people and communities. Five years on, nearly 2,000 communities had engaged with the neighbourhood planning process (Brownill and Bradley, 2017), and about a year later the Wootton and St Helen Without Neighbourhood Plan Steering Group (WSHWNPSG) – the neighbourhood plan for my own locality – was established and joined the throng. During that time, a growing body of literature has emerged, investigating multiple forms of localism and the diverse approaches adopted to neighbourhood planning. Such work has identified numerous critical issues surrounding the localism agenda in general and neighbourhood planning in particular. These issues include questions as to the extent to which power is really devolved to local communities if they remain bound up in pre-existing power structures; the extent to which neighbourhood plans are genuinely community-based if their production relies on a small group of self-nominated residents; the appropriate nature and extent of support provided by LPAs for neighbourhood planning when this is both unspecified in the regulations and variable in practice; and the degree to which community-based plans can really combat the dominance of the growth agenda (Gallent and Robinson, 2013; Wills, 2016; Brownill and Bradley, 2017; Bradley and Brownill, 2017; Bradley et al, 2017; Parker et al, 2017; Vigar et al, 2017; Bradley, 2018; Wargent and Parker, 2018).

Context, aims and audience

This volume contributes to this body of work, engages with many of these issues and develops academic perspectives on neighbourhood planning in some important and distinctive ways. Specifically, although this work is written from an academic and critical perspective, my engagement with the neighbourhood planning process that it draws upon was not as an academic researcher or a consultant advisor. Rather, I participated simply as a member of my local community with a desire to practice my academic discipline – geography – in a live process of place-making in a place in which I was personally invested. Consequently, and in contrast to much existing literature on neighbourhood planning, I address neighbourhood planning not from the perspective of an outsider but as an insider to the process and from a role – vice-chair of the steering group – that was actively and consistently engaged in almost all aspects of the neighbourhood planning process.

Writing from within a neighbourhood planning process enables me to engage in depth with a single case study rather than drawing out commonalities across several neighbourhood plans. This first book-length insider account of neighbourhood planning thereby addresses identified needs for more detailed research into the process by which the content of neighbourhood plans is shaped and the micropolitics and tactics involved in doing so (Parker and Street, 2015; Parker, 2017; Parker et al, 2017), and for closer investigation of the circumstances and ways in which place is constituted and evoked (Bradley, 2018; Lennon and Moore, 2018). Further, by exploring the ways in which we sought to liberate our neighbourhood plan from its official position below and behind the relevant local plan, this volume seeks to challenge the distinctions drawn in planning between the strategic and the local and to unsettle supposedly clear-cut spatial and planning hierarchies by arguing that there is no inherent reason why neighbourhood plans should not be strategic in nature. Along the way and through these endeavours, I also propose conceptual developments by which we might both advance our understanding of the complexities of place-making within neighbourhood planning and interrogate, describe and appreciate the dynamics and contradictions in such place-making (Martin, 2003; Matthews, 2014; Filion, 2014; Brownill, 2017; Etherington and Jones, 2017; Parker and Salter, 2017). As such, while this book is aimed towards an academic audience, it is also written in the hope of providing insights from our own experience to support other people and communities who are developing or revising their own neighbourhood plans. Primarily, however, it is of academic significance, speaking substantively to both the role of neighbourhood planning in the broader planning project and to the production of place in locally contested circumstances, thereby contributing to academic, practitioner and local communities alike.

This is not a claim that our neighbourhood plan is in any way an exemplar of best practice, and this book is not intended to be a 'how to' or 'self-help' guide to neighbourhood planning, partly because I do not think there is a 'one size fits all' approach for an effective neighbourhood plan but also because there already exists a host of practical guidance, including that provided by the government (DCLG, 2012/2018, MHCLG, 2018), dedicated neighbourhood planning websites (e.g. www.neighbourhoodplanner.org.uk and https://neighbourhoodplanning.org), and advice available via LPAs. However, my own experience of neighbourhood planning involves both local specificities and broader or more systemic insights that hold the promise of being more universally relevant or informative than one case study might suggest.

The local specificities include the fact that our neighbourhood plan has been produced jointly between two adjacent parish councils, and that an operational army barracks and former RAF airfield fall entirely within our neighbourhood plan area, spanning the boundary between the two parishes. Although military sites are often excluded from neighbourhood plans, in this instance our 'Designated Area' does include the barracks and airfield because this site was identified for disposal by the Ministry of Defence (MoD) as part of the government review *A Better Defence Estate* (MoD, 2016), reducing the sensitivities of neighbourhood planning for a military site and making the site available for other uses. This site was promptly allocated as a Strategic Development Site by the LPA, an event which occurred in the interim period between the adoption of part one of the local plan and the development of part two of the same plan, constituting a major change in the planning context for our neighbourhood plan. Additional specificities include the proposal within the local plan to delete land from the Green Belt to deliver this proposed development, which local residents vehemently do not want, and a proposal to merge the new development with a nearby settlement, which local residents similarly do not want.

All these features culminated in a complex situation in which we needed to develop a community-led plan that did not contradict the strategic objectives of the local authority but where local residents were strongly against key elements of the proposals for the strategic development site, and in which the major change in development strategy during the evolution of the local plan brought about significant implications for the timing of development of our own neighbourhood plan. Consequently, our experience of neighbourhood planning strikes at the heart of key critical issues for neighbourhood planning, such as the conflict between community-led and state-led planning, between economic growth and sustainable development, between protection of the Green Belt and the release of military sites for housing development in the Green Belt, and between delivering objectives through the plan itself or through the processes associated with developing it.

These key critical issues are themselves among the broader difficulties and shortcomings mentioned earlier, and contribute to other, related, challenges such as how to manage, remove or build constructively upon dissensus within planning, how to carve out and capitalise upon new political opportunities within the planning system, how to deal with the intersection of the multiple and varied spatial and temporal scales associated with different stakeholders and agendas, and how neighbourhood planning might be modified (or overhauled) in the future to deliver good, meaningful or proper localism. Through an interrogation of the specificities of my own neighbourhood planning circumstances and experiences, I work through the broader implications of these for localist agendas and initiatives in their many and varied forms around the world. Similar drives towards localism – tailored to their local social, political, cultural and economic conditions – have been explored in numerous geographical locations, and many of the same challenges and criticisms have come to the fore irrespective of the local conditions within which they are situated. Whether considered in France, Australia, the United States of America, the Netherlands, Italy or Norway, commonalities are not difficult to find. On the one hand, these commonalities relate to the circumstances within which neighbourhood planning takes place, including increasing citizen involvement in hierarchical systems, pressures of austerity measures impacting on LPA capacity, different timescales of significance for communities compared to the state, and the implications of the sustainable development agenda. On the other hand, commonalities can be identified in the problems encountered and critiques exposed, including difficulties in defining the common or public good, tensions between the existing authority at the level of the state and the new autonomy granted at the level of the community, presuppositions of a homogenous community, and the challenge of reconciling local and strategic objectives (Amdam, 2014; Ciaffi, 2014; Dandekar and Main, 2014; van der Pennen and Schreuders, 2014; Bennett, 2017; Burton, 2017; Gardesse and Zetlaoui-Léger, 2017). While acknowledging the unique local conditions within which my own neighbourhood plan has taken shape, then, the challenges encountered and the critical commentary generated in the process have broader-than-local ramifications, in relation to both how neighbourhood planning is done in practice and how we engage academically with neighbourhood planning.

This volume has been written alongside the closing stages of the neighbourhood plan process that it describes (from the pre-submission consultation stage onwards). This is deliberate, to ensure that the insights gained, issues encountered and decisions taken along the way do not become lost or discarded in the event of disappointing results. The success or otherwise of our efforts on the part of our community is only part of the equation. Perhaps more important than the neighbourhood plan as a product is the potential within the neighbourhood planning process for both academic and practitioner audiences. While I have been deeply enmeshed in the neighbourhood planning process for two years now, I maintain a 'helicopter' perspective in this volume, keeping my focus at the level of documents and organisations rather than individuals, and I do my best to avoid slipping into 'us versus them' narratives concerning the relationship between the neighbourhood plan steering group and parish councils on the one hand and the local authority and developers on the other hand. It is also important to confirm that the views and opinions articulated in these pages are entirely my own, except where I explicitly acknowledge that the neighbourhood plan steering group (NPSG) took a position on an issue. Both the NPSG and the two parish councils have had the opportunity to review a draft of this volume, with the purpose of this review process being to check the accuracy of the account that I present rather than to inform or endorse the messages and opinions that I convey. I received one response to this review process, from the Chair of the WSHWNP, who commented as follows:

> It is a detailed document accurately outlining all the issues that we have faced going through the process. You have captured all the relevant details and have pointed to the disconnect between what the government were aiming to do in the Localism Act and how various 'single interest' groups can frustrate and undermine that process, and as I was reading it all the feelings of frustration and disappointment came to the fore. Your comments about the community being denied, excluded and delegitimised are spot on: on a few occasions you say I resented etc – I think you can rightly say we.

Consequently, while this book presents my own views and opinions, the sense of frustration and disappointment that I convey are not unique to me. This review also reinforces the different perspectives and agendas at play in neighbourhood planning, as well as the sense of exclusion on the part of the community that I describe in the chapters that follow. Admittedly, there are occasions when mentioning differences of perspective and raising critical opinions cannot be avoided due to the often contentious politics of space and place involved (Parker et al, 2019), but these should not be seen as 'council bashing' as they simply highlight the key distinctions in perspective that are important to the analytical point being made, and my critical comments are concerned with the system and process rather than people and intentions. I do my best to keep my focus on what we might take from our experience at a systemic or procedural level and in an informative or affirmative fashion, although there is no getting around the fact that this is an unapologetically and uncompromisingly critical reflection on neighbourhood planning from my own insider's perspective.

Analytical themes and chapter outline

By thinking about neighbourhood planning in an academic way from within an individual neighbourhood planning process, several analytical and critical themes emerge concerning place-making and the re-spatialisation of planning, conceptualising publics and public goods, the nature and role of strategy, the significance of scale, performativity and affect in planning practice, gritty questions of power, and biting questions of legitimacy. Taken together, these themes generate different possibilities for the doing of neighbourhood planning. The most significant of these possibilities concerns reconfiguring both the hierarchical relationship between local and neighbourhood plans as documents and the procedural relationship between local and neighbourhood planning as processes, as well as reframing the relationship between the local and the strategic in the context of these reconfigured hierarchical and procedural relationships. In doing so, I propose that we can compose an approach to neighbourhood planning that is more consensual in establishing the terms of engagement while acknowledging and accommodating the antagonistic tendencies inherent to any political process, thereby facilitating an integrative approach to planning that both captures local capacity for strategic thinking while channelling national strategy in a more locally sensitive manner, and promotes a more equal balance between economic, social and environmental agendas. Ultimately, I suggest that only if such drastic modifications are made to how neighbourhood planning functions will its practice be considered legitimate and will it have any longevity. As my argument unfolds, my attention progresses from a critical consideration of the varied ways in which place is constructed in the extant literature through detailed engagement with how place was constructed in the context of my own neighbourhood plan, to an exploration of the implications that these have for our understandings of the making and unmaking of place, our current under-appreciation of the strategic potential of neighbourhood planning, similar under-appreciation of the potential of the affective aspects of neighbourhood planning, and the legitimacy or otherwise of neighbourhood planning now and in the future, as detailed in the following chapter summary.

In Chapter 1, I provide a brief summary of the background to neighbourhood planning in the context of historical trends towards localism in English politics and characterise the aims of localism as they apply to neighbourhood planning. Against this background, I outline the context for the Wootton and St Helen Without Neighbourhood Plan, including local aspirations towards devolution within local government, the impacts of the 'duty to cooperate' between local authorities, and the implications of this for the Green Belt in my locality. Subsequently, I describe the process, timing and progress of the WSHWNP, paying particular attention to its relationship with the relevant local plan.

Chapter 2 engages with different conceptualisations of space as a starting point for the detailed analysis through the subsequent chapters of different ways in which place was constructed during our neighbourhood planning experience. Critiquing the common tendency to conceptualise place in binary terms – insider/outsider, site/place, abstract/lived – in planning literatures (e.g. Fenster and Misgav, 2014; van der Pennen and Schreuders, 2014; Bradley, 2017, 2018) and building upon previous recognition that vertical and horizontal spatial imaginaries of power in planning are overly simplistic (Brownill, 2017), I explore vertical, horizontal and networked spatial imaginaries and establish the ways in which our neighbourhood plan area is simultaneously all three, in that it is bound to the vertical, contested by the horizontal and driven by the network. In doing so, I set the scene for fuller discussion of the ways in which this configuration remains insufficient as it takes us to the significance of – but does not directly address – the active way in which the different stakeholders in neighbourhood planning drive the network in the construction of place and the implications of that for place identity and attachment. It is the opportunity to interrogate these issues in depth through thorough engagement with a single neighbourhood plan process from the inside, which this volume delivers. The subsequent three chapters address different perspectives on this more active consideration of place in neighbourhood planning: thematic, strategic and performative.

In Chapter 3, I explore the different ways in which three themes are understood by different stakeholders in the neighbourhood planning process – the environment, the Green Belt and sustainable development – and the implications that this brings for the effective alignment of a neighbourhood plan with a local plan and for the effectiveness of a neighbourhood plan in delivering its aims. Contributing to an identified need for greater appreciation of the dynamics and contradictions of place-making (Brownill, 2017) and questions concerning the degree to which neighbourhood plans resolve tensions between contradictory planning agendas (Brownill and Bradley, 2017), this discussion questions the extent to which different stakeholders in neighbourhood planning share a common symbolic space (Mouffe, 2005; Kenis, 2018) by extending the idea of parochialism beyond the realm of spatial proximity to the realm of fields of practice (Madanipour and Davoudi, 2015). In doing so, it highlights that such tensions are unlikely to be resolved by neighbourhood planning because in true post-political fashion such divergent underlying understandings are elided rather than explicitly addressed (Haughton et al, 2013; Kenis, 2018) in the planning process.

Chapter 4 engages with concerns that it can be hard to articulate local issues in terms of strategy or principle (Gallent and Robinson, 2013) and considers three ways in which we tried to adopt a strategic approach: questioning what 'strategic' really means in the context of planning policies, developing our own spatial strategy for the content of the neighbourhood plan, and developing our strategy for the way in which the timing of our neighbourhood plan interfaced with the timing of the emerging local plan with which it had to conform. Through this discussion, the importance of moving away from considering neighbourhood plans as finished documents to consider the interaction of the processes of production of plans at different levels in optimising the effectiveness of neighbourhood plans is highlighted (Parker and Salter, 2017; Vigar et al, 2017; Bradley, 2018). Further, the difficulties encountered in doing this in the face of entrenched professional views as to the appropriate place and time of a neighbourhood plan within the planning system and the assumption formalised in the planning regulations that strategy is a feature of a spatial scale rather than a practice or approach that is independent of scale are excavated. This takes the debate beyond concerns over the appropriateness, possibility and implications of demanding that local knowledge be translated into strategic planning speak (Parker et al, 2015; Wargent and Parker, 2018) to concerns over our very understanding of what it means to be strategic and how the strategic should relate to the local, prompting calls for a reconfiguration of how neighbourhood plans might fit into the broader planning framework.

In Chapter 5, I focus on three ways in which neighbourhood planning can be formally understood as performative: political, practical and presentational. Contributing to the existing literature by providing further detail on the processes by which the content of neighbourhood plans is shaped (Parker, 2017; Parker and Street, 2015; Vigar et al, 2017; Wargent and Parker, 2018), I reinforce concerns about the boundedness of neighbourhood planning to pre-existing political structures and power dynamics, thereby reducing their potential to deliver genuine local autonomy through the imposition of formal hierarchies of power to control the activities of a neighbourhood planning team. Subsequently, I add to calls for greater clarity as to the LPA's duty to support neighbourhood planning by highlighting both the capacity for the imposition of softer forms of power through the practice of policy modulation or by acting as if a local plan had already been adopted, and the difficulties encountered in catering for multiple audiences and communicating community ownership of the neighbourhood plan. This emphasises again the need to think beyond the document itself to consider related processes for other aspects of the planning system and the significance of presentational factors.

Chapter 6 pulls together the emergent analytical and critical threads of these chapters to work through their implications with regard to place attachment and ways in which we might conceptualise spatial experiences of neighbourhood planning, the legitimacy of neighbourhood planning, and its future prospects. I draw attention to the potentially detrimental impact on place attachment arising from participation in neighbourhood planning as a counterfoil to assumptions of place attachment as a bottomless pit of positive affect that can be enhanced through neighbourhood planning, and explore different conceptualisations of threshold experiences (hybrid, liminal and chimerical) by which we might understand these experiences. I also present a scathing account of the legitimacy of neighbourhood planning from the perspective of my own experience, and I argue that the prospects for neighbourhood planning are bleak unless substantive and radical modifications are made not just to the process of neighbourhood planning itself but to the broader planning project within which it sits.

Finally, the conclusion draws together the outcomes of the preceding discussions to highlight how the specifics of one neighbourhood planning process can inform both the academic understanding and practical development of neighbourhood planning and can be of relevance both within and beyond the English case study. In particular, a series of nested insights is articulated through which the core argument about the need for radical reformulation of how neighbourhood planning is structured and delivered is established. The thematic emphases highlighted here (e.g. power, relationality, context, creativity and so on) hold relevance across varied circumstances and instances of localist initiatives, enabling this volume to contribute to localism beyond the English context. This also serves to reinforce the call to reconfigure the relationship between the strategic and the local, between neighbourhood and local planning, and proposes that although the implications of neighbourhood planning for place attachment have the potential to be negative, the conceptualisation of such experiences as outlined in Chapter 6 might yet bring opportunities for both oppositional tactics *within* neighbourhood planning and strategic reconfiguration *of* neighbourhood planning.

Two further critical points emerge from the concluding discussion. The first highlights the value in thinking about planning from beyond planning, such that by unsettling conventional assumptions (such as the relationship between local and neighbourhood planning) we might prompt creative solutions, and by intervening in planning and academic literature about planning from the conceptual perspective of different disciplines (such as the application of geographical spatial concepts to the practice of planning) we might identify new avenues for both practical and critical interrogation. The second draws together the shortcomings of neighbourhood planning previously identified in the literature, their development into a more substantive critique of the legitimacy of neighbourhood planning in this volume, and concerns over the potentially detrimental impact of neighbourhood planning on place attachment, to propose that what is needed at this juncture is an explicitly localist strand of critical scholarship. I suggest that the severity and significance of the failings of neighbourhood planning are sufficient to determine that neighbourhood planning cannot possibly deliver its stated aims, nor can it achieve substantive outcomes for communities without either relying on higher levels of the planning system – which undermines the very ethos of localism – or radical modification of the way in which neighbourhood planning is done, and as such cannot be considered legitimate. This, combined with the potential erosion of the very place identity and place attachment upon which localism – and especially neighbourhood planning – depends has dire implications as it suggests that despite proclamations to the contrary, neighbourhood planning might progressively destroy rather than promote the local and dismantle rather than bolster participatory democracy.

Conclusion

Tinkering with the relationship between local and neighbourhood planning within the confines of neighbourhood planning is a step in the right direction, but in and of itself is insufficient. Without requisite amendments to the inbuilt prioritisation of economic growth over environmental and social concerns within the NPPF (Bradley, 2018), which not only renders neighbourhood planning null and void by effectively silencing local voices but also renders the very economic growth that it promotes inherently unsustainable by denying the reliance of economic growth on both society and the environment, such modifications will, I fear, amount to little more than rearranging the deckchairs on the Titanic. There needs to be a significant shift away from economic growth at all costs towards properly sustainable development, which genuinely balances economic, social and environmental needs, and which incorporates the views, aspirations and concerns of residents in determining that balance in a strategic rather than reactive fashion. To adopt a local perspective is not the same as NIMBYism, and it is inappropriate for local communities to be accused of NIMBYism when they not only support development in their area and are quite capable of thinking and acting strategically but are prevented from doing so by the very legislation and planning system that claims to value and support local involvement in planning. It is equally unacceptable to accuse local communities of being parochial and reactive when that is the only type of response that the planning system – despite claims to the contrary – allows them to express.

We need to reconfigure neighbourhood planning within the broader planning project, bringing planning back into the realms of the political, facilitating strategic thinking at the neighbourhood level and sensitivity to locality at the local level, and formalising a mechanism to facilitate the delivery of more collaborative planning deliberations and more equitable planning outcomes. The future of neighbourhood planning could be very bright but in its current configuration it is deeply troubling, and the prospects for neighbourhood planning as I see them can only be reversed by accepting that neighbourhood planning must be permitted to do that which it is not currently permitted to do; by being that which it is not currently permitted to be. The formalised exclusion of local people from strategic issues, the lack of parity between environmental, economic and social concerns, and the proceduralised dismissal of local place knowledge within the planning system merely serve to underline how powerful and productive local communities, their knowledge and their energy could be in more permissive, more progressive circumstances.

References

Amdam, J (2014) Flexible local planning: linking community initiative with municipal planning in Volda, Norway. In N Gallent and D Ciaffi (eds) *Community action and planning: contexts, drivers and outcomes*. Policy Press: Bristol, p281–299

Bennett, L (2017) The many lives of neighbourhood planning in the United States: much ado about something? In S Brownill and Q Bradley (eds) *Localism and neighbourhood planning: power to the people?* Policy Press: Bristol, p231–247

Bradley, Q (2017) A passion for place: the emotional identifications and empowerment of neighbourhood planning. In S Brownill and Q Bradley (eds) *Localism and neighbourhood planning: power to the people?* Policy Press: Bristol, p163–179

Bradley, Q (2018) Neighbourhood planning and the production of spatial knowledge. *The Town Planning Review* 89 (1): 23–42

Bradley, Q and Brownill, S (2017) Reflections of neighbourhood planning: towards a progressive localism. In S Brownill and Q Bradley (eds) *Localism and neighbourhood planning: power to the people?* Policy Press: Bristol, p251–267

Bradley, Q, Burnett, A and Sparling, W (2017) Neighbourhood planning and the spatial practices of localism. In S Brownill and Q Bradley (eds) *Localism and neighbourhood planning: power to the people?* Policy Press: Bristol, p57–74

Brownill, S (2017) Assembling neighbourhoods: topologies of power and the reshaping of planning. In S Brownill and Q Bradley (eds) *Localism and neighbourhood planning: power to the people?* Policy Press: Bristol, p145–161

Brownill, S and Bradley, Q (2017) Introduction. In S Brownill and Q Bradley (eds) *Localism and neighbourhood planning: power to the people?* Policy Press: Bristol, p1–15

Burton, P (2017) Localism and neighbourhood planning in Australian public policy and governance. In S Brownill and Q Bradley (eds) *Localism and neighbourhood planning: power to the people?* Policy Press: Bristol, p215–230

Ciaffi, D (2014) Active communities of interest and the political process in Italy. In N Gallent and D Ciaffi (eds) *Community action and planning: contexts, drivers and outcomes*. Policy Press: Bristol, p201–215

Dandekar, H and Main, K (2014) Small-town comprehensive planning in California: medial pathways to community participation. In N Gallent and D Ciaffi (eds) *Community action and planning: contexts, drivers and outcomes*. Policy Press: Bristol, p157–176

DCLG (2012/2018) *Neighbourhood planning (general) regulations* www.legislation.gov.uk/uksi/2012/637/contents/made

Etherington, D and Jones, M (2017) Re-stating the post-political: depoliticization, social inequalities, and city-region growth. *Environment and Planning A: Economy and Space* 50 (1): 51–72

Fenster, T and Misgav, C (2014) Memory and place in participatory planning. *Planning Theory and Practice* 15 (3): 349–369

Filion, P (2014) The scaling of planning: administrative levels and neighbourhood mobilisation as obstacles to planning consensus. In N Gallent and D Ciaffi (eds) *Community action and planning: contexts, drivers and outcomes*. Policy Press: Bristol, p261–279

Gallent, N and Robinson, S (2013) *Neighbourhood planning: communities, networks and governance*. Policy Press: Bristol

Gardesse, C and Zetlaoui-Léger, J (2017) Citizen participation: an essential lever for urban transformation in France? In S Brownill and Q Bradley (eds) *Localism and neighbourhood planning: power to the people?* Policy Press: Bristol, p199–214

Haughton, G, Allmendinger, P and Oosterlynck, S (2013) Spaces of neoliberal experimentation: soft spaces, postpolitics, and neoliberal governmentality. *Environment and Planning A* 45: 217–234 https://neighbourhoodplanning.org

Kenis, A (2018) Post-politics contested: why multiple voices on climate change do not equal politicisation. *Environment and Planning C: Politics and Space* DOI:10.1177/0263774X18807209

Lennon, M and Moore, D (2018) Planning, 'politics' and the production of space: the formulation and application of a framework for examining the micropolitics of community place-making. *Journal of Environmental Policy and Planning* DOI:10.1080/1523908X.2018.1508336

Madanipour, A and Davoudi, S (2015) Localism: institutions, territories, representations. In S Davoudi and A Madanipour (eds) *Reconsidering localism*. Routledge: New York and London, chapter 2. (no page numbers) Accessed 6 Dec 2018

Martin, DG (2003) 'Place-framing' as place-making: constituting a neighbourhood for organizing and activism. *Annals of the Association of American Geographers* 93 (3): 730–750

Matthews, P (2014) Time, belonging and development: a challenge for participation and research. In N Gallent and D Ciaffi (eds) *Community action and planning: contexts, drivers and outcomes*. Policy Press: Bristol, p41–56

MHCLG (2018) *Guidance: neighbourhood planning* www.gov.uk/guidance/neighbourhood-planning-2

Milne, R (2016) Residents reject Derbyshire neighbourhood plan. *Planning portal* www.planningportal.co.uk, news article 434 Accessed 10 Jan 2019

MoD (2016) *A better defence estate* www.gov.uk/government/publications/better-defence-estate-strategy

Mouffe, C (2005) *On the political*. Routledge: Abingdon

Parker, G (2017) The uneven geographies of neighbourhood planning in England. In S Brownill and Q Bradley (eds) *Localism and neighbourhood planning: power to the people?* Policy Press: Bristol, p75–91

Parker, G, Lynn, T and Wargent, M (2015) Sticking to the script? The co-production of neighbourhood planning in England. *The Town Planning Review* 86 (5): 519–536

Parker, G, Lynn, T and Wargent, M (2017) Contestation and conservatism in neighbourhood planning in England: reconciling agonism and collaboration? *Planning Theory and Practice* 18 (3): 446–465

Parker, G and Salter, K (2017) Taking stock of neighbourhood planning 2011–2016. *Planning Practice and Research* 32 (4): 478–490

Parker, G, Salter, K and Wargent, M (2019) *Neighbourhood planning in practice*. Lund Humphries: London

Parker, G and Street, E (2015) Planning at the neighbourhood scale: localism, dialogical politics, and the modulation of community action. *Environment and Planning C: Government and Policy* 33: 794–810

van der Pennen, T and Schreuders, H (2014) The fourth way of active citizenship: case studies from the Netherlands. In N Gallent and D Ciaffi (eds) *Community action and planning: contexts, drivers and outcomes*. Policy Press: Bristol, p106–122

Vigar, G, Gunn, S and Brookes, E (2017) Governing our neighbours: participation and conflict in neighbourhood planning. *The Town Planning Review* 88 (4): 423–442

Wargent, M and Parker, G (2018) Re-imagining neighbourhood governance: the future of neighbourhood planning in England. *The Town Planning Review* 89 (4): 379–402

Wills, J (2016) *Locating localism: statecraft, citizenship and democracy*. Policy Press: Bristol www.neighbourhoodplanner.org.uk

1 Localism and neighbourhood planning

Introduction

Set against a long history of tension and fluctuation between localising and nationalising tendencies in English democracy, localism is the latest stage in this unfolding dynamic. As one specific example of localism, neighbourhood planning enables local communities to produce their own development plan for their immediate neighbourhood, which is deemed to grant them a degree of control over what type of development takes place, and where, in their area. Taking slightly different forms in parished and unparished areas, my focus here is on parished areas simply because my own experience of neighbourhood planning has been in such an area, but the substantive points arising from this focus are applicable to unparished areas as well. In this chapter, I provide brief background to localism in English politics and to neighbourhood plans as one example of this agenda, along with an overview of their affordances and constraints in terms of what they can and cannot cover and a summary of some of the key criticisms widely publicised in the existing literature. Subsequently, I describe in summary form the context for the development of my own neighbourhood plan, taking into consideration the broader changes to local government arrangements made possible by the Localism Act, the ways in which local authorities in my area have responded to these, and the specific context of the Wootton and St Helen Without Neighbourhood Plan (WSHWNP). Finally, I summarise the approach, process and timescale for our neighbourhood plan, and detail the specific circumstances that make our experience such an informative case study, especially in relation to the presence of a sizeable military site within our area that was anticipated for disposal by the MoD, and the respective timings of our neighbourhood plan and the relevant local plan, whereby our neighbourhood plan was developed after the adoption of the first part of the local plan but before the adoption of the second part of the same local plan. In essence, this chapter provides the local government, neighbourhood planning and geographical context for the substantive analysis delivered in Chapters 3 to 6.

Localism

Participatory and lower level (more local) forms of governance are now promoted by all three mainstream political parties in the UK and localism has been promoted as an alternative to centralised forms of governance since the Conservative-Liberal Democrat Coalition government came to power in 2010 but it is not an entirely new phenomenon, with a longer history and more diverse approaches than are sometimes assumed (Owen et al, 2007; Gallent and Robinson, 2013; Davoudi and Madanipour, 2015a; Wills, 2016a, 2016b). A tension between standardisation through centralist approaches and responsiveness through localist approaches to government has been identified stretching as far back as the emergence of local government within the centralisation of state government following the Norman Conquest (Wills, 2016a). More recently, though, the move away from centralised control can be traced through the Thatcherite roll-back of the state in favour of marketisation in the 1980s, Labour's dualistic approach of setting national standards and performance targets on the one hand but allowing local freedoms in return for good performance on the other hand, and the Coalition government's establishment of local authorities' power to act in the interests of their communities and reduction of regional directives (Gallent and Robinson, 2013).

In this progressive shift from government to governance, central government adopted a more permissive and facilitative role, reforming the planning process to involve diverse stakeholders with different perspectives and agendas (Gallent and Robinson, 2013; Wills, 2016a; Brownill, 2017, Parker, 2017). Exemplified in the Localism Act 2011 (DCLG, 2011), the aims were to place decision-making in the hands of those affected by the decisions to be made and democratise policymaking beyond professional planners and politicians. Six concrete actions were identified for all aspects of government to return power to local people, including reducing bureaucracy, opening government to public scrutiny, increasing local accountability and empowering communities to act on their own concerns (Wills, 2016a).

The effects of localism can be seen across society, with power increasingly transferred from central government to local government, including through the establishment of new structures of governance such as city-regions and new opportunities for establishing combined or unitary authorities (DCLG, 2016a). Power has also been transferred to newly-established commissioning arrangements for local policing and health provision, and to local communities in the form of their ability to establish free schools as part of the effort to create a Big Society (Wills, 2016a; Brownill, 2017), leading to debates as to whether such moves constitute localised choice or a postcode lottery (Wills, 2016a). Welfare and wellbeing have also been transformed through the localism agenda as 'care in the community' came into effect with the roll-back of the welfare state, imposing greater individual responsibility on citizens for their own health and wellbeing in the face of increasing costs, state failures and the need for more active citizenship (Milligan and Conradson, 2006).

Neighbourhood planning is simply one more example of the impacts of localism, incorporating residents and communities into the formal machinery of the planning framework. In England, this development builds upon earlier attempts to engage a broader population in the co-ordination of service delivery, such as Local Strategic Partnerships, Local Area Agreements, Community Strategies, and parish and community plans (Gallent and Robinson, 2013; Wills, 2016b; Vigar et al, 2017) but this current instantiation of 'community engagement' has statutory weight (Gallent and Robinson, 2013; Bradley et al, 2017). Neighbourhood planning establishes a new tier in the planning framework, bringing new rights for civic engagement in planning through the development of community-led plans in a community rather than a state-led form of localism (Wills, 2016a; Brownill and Bradley, 2017). As a form of localism, the aims of neighbourhood planning have been described as enhancing community empowerment, encouraging community acceptance of growth, and remaking the relationship between the public and planning as planning becomes collaborative and non-professional through the involvement of the public in its practice (Davoudi and Madanipour, 2015b; Brownill, 2017).

Noting their formal title as Neighbourhood *Development* Plans highlights their emphasis on development as well as neighbourhood and planning. While one aim is to engage neighbourhoods in decision-making and another is to modify the nature of planning, equally important is the intention of neighbourhood planning to encourage local acceptance of – and discourage appeals against – development proposals, thereby increasing and accelerating economic growth (Gallent and Robinson, 2013; Bradley et al, 2017). To this end, financial incentives are made available in the form of a greater proportion (25% rather than 15%) of the Community Infrastructure Levy (the funds secured from developers to help deliver infrastructure) for communities that have a neighbourhood plan in place compared to those that do not (Brownill, 2017), although there is little scope to do any more than state local preferences for what that funding is allocated to deliver (Bradley et al, 2017).

So, what are neighbourhood development plans (hereafter neighbourhood plans), and what are local communities able to achieve and are precluded from achieving by the introduction of this new tier to the planning system? Neighbourhood plans are planning policy documents that – once adopted – sit alongside the local plan for an area and help to determine how planning applications affecting the area covered by the neighbourhood plan are to be determined. In other words, neighbourhood plans can help to shape the location, scale, type and style of development in a local area. In terms of location, neighbourhood plans are able – but are not obliged – to allocate additional sites for development but can also designate specific sites and features as either Local Green Spaces or Heritage Assets to protect those sites and features from development pressure. In terms of scale, although neighbourhood plans cannot refuse strategic development allocations, they can specify maximum sizes for individual developments in terms of the number of dwellings allowed and many do take this step as a means of encouraging a model of development that favours smaller-scale developments over large, speculative developments (Bradley et al, 2017; Bradley and Brownill, 2017). In terms of development type, neighbourhood plans can encourage higher-than-minimum standards for issues such as affordable housing, sustainable design and environmental protection, and can encourage the development of community infrastructure or transport improvements. In terms of style, neighbourhood plans can specify requirements for building heights, plot layouts, boundary treatments and architectural appearance. However, the inability to refuse strategic development allocations is a major drawback for communities as it forces compliance (and – one might argue – complicity) with precisely the type of large-scale development that communities most often do not want. In addition, while neighbourhood plans can encourage higher than minimum standards in certain regards, encouragement is all they can now do as the Deregulation Act 2015 (Cabinet Office, 2015) prevented neighbourhood plans from establishing or enforcing additional standards for technical and environmental aspects of development (Bradley et al, 2017). This is also a significant drawback as it links to the perceived problems of large-scale developments, and enhancing sustainability is often a key driver for the development of a neighbourhood plan in the first place.

As a planning policy document, neighbourhood plans are also constrained in terms of their ability to address issues and concerns of local communities that fall outside the remit of planning (Bradley et al, 2017), such as traffic calming measures, installing additional dog bins or benches, maintaining verges and facilities, improving pavements for those with mobility impairments, and so on. While it is often the case that the neighbourhood planning process leads to the identification of a range of community concerns and projects that the parish council or community forum can then take forwards, the neighbourhood plan itself cannot address these issues and there is no additional funding for their delivery, so the danger is that expectations are raised and cannot be met, risking disengagement from rather than enhanced engagement with the idea of active citizenship (Gallent and Robinson, 2013; Wills, 2016a). This is compounded by the cessation of steering group or community forum involvement in the neighbourhood plan once it has been made, as responsibility for its implementation in relation to planning applications lies primarily with the local authority (Wargent and Parker, 2018). Unless the team behind the neighbourhood plan can establish and sustain a delivery structure and secure funds for the delivery of community projects, it is unlikely that such projects will come to fruition (Wills, 2016a). While on the one hand, the cessation of involvement on the part of the volunteers behind a neighbourhood plan might encourage involvement in the first place as it is seen as a fixed-term rather than permanent dedication of time and effort, it also risks condemning neighbourhood planning to a very fleeting experiment in localism given the effort that is involved in establishing the steering group or community forum necessary to develop or revise a neighbourhood plan. In addition, although a neighbourhood plan carries equal weight – once made – to the non-strategic aspects of the local plan, any decisions made on planning applications will involve a degree of judgement on the part of the local authority and to that extent remain somewhat discretionary (Brownill, 2017).

While it has been proposed that there are very few requirements stipulated as to what a neighbourhood plan needs to contain (Wills, 2016a), it is perhaps more informative to consider the controls on what a neighbourhood plan cannot or should not contain, as this is the greater source of constraint on the effectiveness of neighbourhood planning. We have already seen that communities cannot refuse strategic housing allocations, cannot address non-planning issues, cannot determine the use of the additional funds received through the Community Infrastructure Levy, and cannot stipulate higher than minimum technical and environmental standards, but neighbourhood plans should also not duplicate policies that are already enshrined in the National Planning Policy Framework or the relevant local plan. While superficially this seems sensible, it is problematic in four main ways:

1 It flies in the face of the expectation that neighbourhood plans should be community-led if their content is defined in relation to higher level planning policies.
2 It denies communities the ability and right to reinforce their own genuine commitment to higher-level strategic policies.
3 It neglects the role of a neighbourhood plan in communicating planning policies and priorities to local residents who are likely to be unfamiliar with both the National Planning Policy Framework (NPPF) (2012/2018) and the local plan.
4 If the NPPF or local plan policy priorities are actively and strongly supported locally and that support is not evidenced in the neighbourhood plan, the bodies behind the neighbourhood plan (steering groups, community forums and parish councils) are put in a position of being perceived as either not listening to their communities or simply ignoring them, potentially contributing to a break down in social relations rather than a stronger civil society.

Ironically, then, neighbourhood planning teams are obliged to provide explicit support for development allocations that their communities might rather not have in their area while they are simultaneously prevented from explicitly articulating their support for strategic policies that their communities do genuinely support. This issue is partially resolved by the requirement not to include *unnecessary* duplication of policies (DCLG, 2012), but all this really does is inject potential for conflict in the determination of what counts as necessary duplication. Consequently, a neighbourhood plan is severely limited in what it can achieve as it must confine itself to matters that can be addressed through planning; it cannot refuse the very type of development that locals are most likely to dislike; it can do no more than encourage higher than minimum standards in some of the aspects of planning that are most important locally; there is no mechanism to support the ongoing engagement with and delivery of community concerns on the part of those originally involved in the development of the plan; genuine support for higher level policies cannot be articulated; and even once made the judgement required in determining planning applications might render the neighbourhood plan redundant.

In addition to these 'in principle' constraints, the most significant criticisms of the neighbourhood planning process identified in research into how it has been practised include:

1 The enforced compliance with strategic development allocations, as outlined earlier (Gallent and Robinson, 2013; Wills, 2016a; Brownill and Bradley, 2017; Bradley and Brownill, 2017).

2 Issues over the self-selection of volunteers, leading to over-representation on steering groups of those who are wealthy, vocal, time-rich and highly educated, leading to the reinforcement of pre-existing structural inequalities (Gallent and Robinson, 2013; Parker, 2014; Wills, 2016a, 2016b; Brownill and Bradley, 2017; Bradley and Brownill, 2017; Colomb, 2017).
3 Inherent power imbalances between the local community representatives, the local planning authority and any developers or site promoters that might be involved, reducing the potential of neighbourhood planning to resolve tensions, all of which is reinforced by the extensive need for expertise in the development of a neighbourhood plan, given the requirements for a robust evidence base, the drafting of policies that are operable in legal terms, and the assessment of the emerging plan for environmental impacts (Gallent and Robinson, 2013; Bradley and Brownill, 2017; Bradley, 2018; Brownill and Bradley, 2017; Parker, 2017; Wargent and Parker, 2018).
4 The assumption implicit within neighbourhood planning that neighbourhoods are self-contained, homogeneous, consensual and easy to define (Brownill, 2017; Colomb, 2017; Parker, 2017).
5 Persistent conflict between the use values of the local community and the exchange values of the promoters of development, combined with a lack of genuine or authentic dialogue between the local community and the local authority, both of which limit the potential for effective collaboration (Parker, 2012; Gallent and Robinson, 2013; Wills, 2016a; Bradley et al, 2017; Wargent and Parker, 2018).
6 The lengthy, time-consuming and onerous nature of the process, with the average time to produce a neighbourhood plan being 27 months (Gallent and Robinson, 2013; Brownill, 2017; Parker, 2017; Wargent and Parker, 2018), which limits the number of people likely to consider themselves either capable of or willing to become and remain involved in neighbourhood planning.

The evidence seems to suggest, then, that far from representing or instantiating a radical – or even modest – shift in power dynamics within the planning system, as some have suggested, the rhetoric of neighbourhood planning does not match the reality (Gallent and Robinson, 2013). Rather than heralding the dawn of a brave new world of community-led collaborative planning, I would argue that neighbourhood planning can be seen as little more than a thinly veiled strategy to remove any remaining vestige of democratic resistance from the planning process.

Such strong sentiment raises the question as to why on earth I became involved in my local neighbourhood plan. One reason is that I had not done any academic reading about neighbourhood planning until well into the process, as my reasons for becoming involved were personal rather than academic. I volunteered because as a geographer, I am interested in place-making and the neighbourhood plan presented an opportunity to explore this from inside a process of place-making in relation to a place in which I was personally invested. It had also been a while since I had undertaken any significant voluntary work and the estimated time demands of monthly meetings seemed feasible alongside my formal employment. I was also attracted by the opportunity it brought for me to engage more directly with my former career in local government and training in environmental management than I had been able to in my academic career to date. As it turned out, the anticipated time demands were massively underestimated, my role swiftly shifted from steering to full-on delivery of the neighbourhood plan, and the circumstances within which we were developing our plan altered radically before we had even had a chance to get started, which raised the stakes considerably, as detailed next.

Neighbourhood plan context

The County of Oxfordshire is largely rural, with a significant swathe of the county designated as Green Belt around the city of Oxford, in which development is not permitted except in exceptional circumstances. However, it also has one of the strongest economies in the South East of England, characterised by education and research, tourism, science and technology, car manufacturing, and health. Within the city of Oxford, the strong and growing economy, Green Belt constraints on development at its edges, accessibility to London, and large student population of two universities make Oxford one of the least affordable cities in the UK.

Oxfordshire has a two-tier local government structure, made up of Oxford City Council, four rural district councils and Oxfordshire County Council. Within this context, the Wootton and St Helen Without Neighbourhood Plan area falls within the jurisdiction of the Vale of White Horse District Council (VWHDC), just to the south west of the city of Oxford, and just north of the market town of Abingdon.

The city, district and county councils have somewhat complicated inter-authority relations as, on the one hand, they increasingly work together on cross-border issues and in pursuit of economic growth but, on the other hand, have different views and priorities on certain issues. With the neoliberalisation of local government has come a widening of participation to include the private sector and the voluntary sector, which has given rise to a strong emphasis on cross-boundary co-operation in pursuit of economic development. Through the Oxfordshire Growth Board, the local authorities work in partnership to boost business growth that is focused upon Oxford as the identified centre around which wider activity revolves, effectively integrating two forms of voluntary devolution: a city-region (a county-area that is functionally oriented to a specific city) and an enterprise partnership (business-led partnerships to promote economic growth across a functional economic area). Despite such examples of increasing co-operation, though, tensions persist between different local authorities within the county, for example in relation to the Green Belt and the potential for a reconfiguration of local government. Facing significant growth in both its population and its economy in the coming years, existing problems of homelessness and inequality, and an insufficient supply of land within the city to meet its forecast housing need, Oxford City Council is under pressure to expand at its edges. However, as this land falls within neighbouring districts rather than within its own boundaries, such development can only take place with the support of the surrounding district councils, but as the Green Belt is highly valued by residents, such support would be unpopular among the district councils' own electorates. Similarly, different local authorities have different aspirations when it comes to potential local government restructures. The devolution agenda allowed for the establishment of combined authorities, geared towards economic development, regeneration and transport and preferably with an elected mayor, and re-opened the door to debates about the potential for a unitary authority in Oxfordshire. However, different authorities have different preferences for local government structure, which often pitches the districts against the county.

However, the devolution agenda has introduced other requirements on the part of local authorities that bring the potential to force – or at least encourage – further co-operation on key strategic issues. The Localism Act 2011 also introduced the Duty to Cooperate whereby local authorities and specified agencies (for example, the Environment Agency, Natural England, Highways England and Historic England) are expected to cooperate with one another in a constructive, active and ongoing way on strategic or cross-boundary issues such as housing, job markets, transport and the environment (DCLG, 2011). That said, local authorities are not obliged to cooperate on such matters if doing so would contravene other aspects of national planning policy. Thus, while the Duty to Cooperate might encourage district councils in Oxfordshire to provide land around the edges of the city for housing development, they are not obliged to do so on the grounds that it would contravene Green Belt protection. In practice, the district councils have agreed to accept some of the unmet housing need from Oxford, but it is for each district council to decide how and where it provides for that development, which is conducted through the local planning process.

The relevant local plan for the Wootton and St Helen Without Neighbourhood Plan is that for the Vale of White Horse District Council. This plan is being produced in two parts, the first of which established the strategic approach to development across the district and was completed in 2016 (VWHDC, 2016), and the second of which stipulates detailed policies and additional sites and was due for adoption in Spring 2019 (VWHDC, 2018). The allocation of additional development sites in the Local Plan Part 2 (LPP2) is especially pertinent to the WSHWNP as, in the interval between the adoption of Part 1 and the release of the consultation draft of Part 2, a significant change in circumstances occurred locally. The Ministry of Defence announced its intention to dispose of several former RAF airfields, ostensibly under the guise of enhancing and streamlining the defence establishment (MoD, 2016), making these sites available for development. This coincided with the release of a call for proposals from local authorities for government funding to support the delivery of a series of garden villages, towns and cities (DCLG, 2016b), and – predictably – a number of former airfields have been proposed as sites for the development of garden villages whether or not they are officially supported by the national scheme.

The significance of these developments lies in the presence of a former RAF barracks and airfield – which is currently occupied by two army regiments – within the area covered by our neighbourhood plan. Dalton Barracks and Abingdon Airfield span the boundary between the two parishes covered by the plan and the site was identified as a Strategic Development Site within LPP2 (see Figure 1.1). This

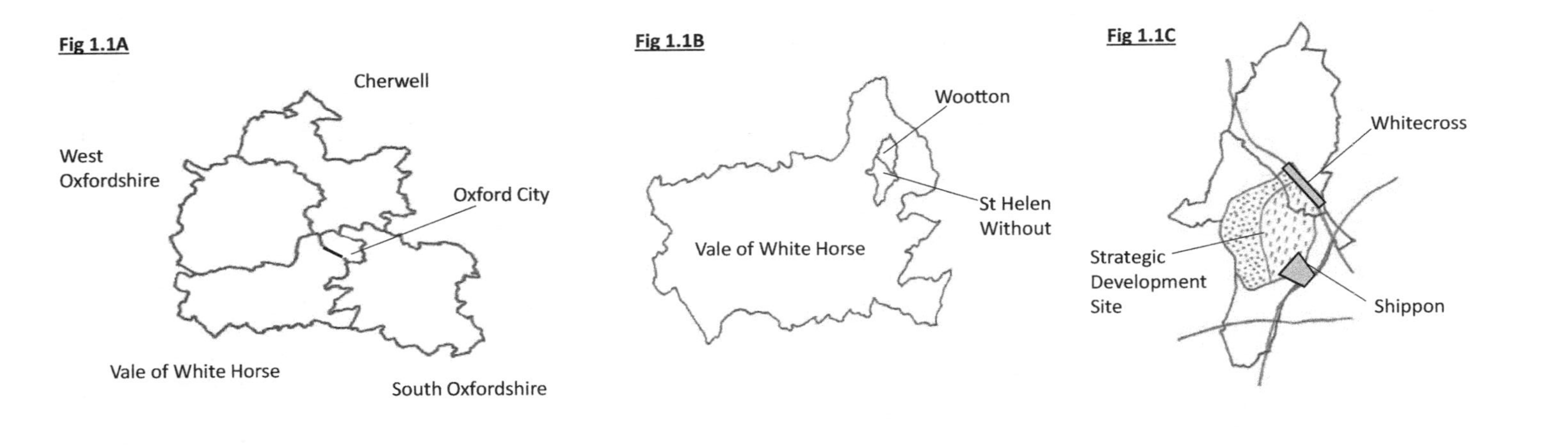

Legend:

1.1A Local authority areas within Oxfordshire

1.1B Wootton and St Helen Without Parishes in the Vale of White Horse DC area

1.1C Initial proposal for Strategic Development Site within our Designated Area

Figure 1.1 Locating the WSHW Neighbourhood Plan

Source: Author

represented a radical shift in the development proposals for the two parishes of Wootton and St Helen Without as, in contrast to the multiple smaller sites identified for development in Part 1 of the local plan, Part 2 allocated all proposed development for the parishes to that one new strategic site at Dalton Barracks. It is also as part of this development that the VWHDC intends to provide its share of the unmet housing need in Oxford under the Duty to Cooperate. The formal allocation within the period of the local plan (up to 2031) is 1,200 dwellings to help provide for Oxford's unmet housing need, but a longer-term development of up to 4,500 dwellings was suggested in early drafts of LPP2.

This major new allocation for development, though, is just one of several significant development pressures either within the two parishes covered by the WSHWNP or in sufficiently close proximity to them to be a cause for concern as to their impacts on local communities, infrastructure and environment. Other examples include proposals to develop a new reservoir, for which the land safeguarded (set aside for later use) creeps into the southern end of the parish of St Helen Without; two new Park and Ride sites, which Oxford City Council seeks to relocate from inside the city boundary to land beyond its boundaries; and the proposed 'expressway' between Oxford and Cambridge, potential corridors for which traverse our Designated Area (the area covered by the neighbourhood plan).

It is within this combination of contexts that the WSHWNP has been developed. As detailed in the neighbourhood plan (WSHWNPSG, 2018), although close to Oxford, the two parishes are rural in nature, with a scattering of small settlements nestling in an agricultural area, the vast majority of which is designated as Green Belt which both reflects and protects the area's agricultural heritage. Demographically, the area is less ethnically diverse than the national average, and we have a higher than average proportion of residents living in communal establishments, almost entirely due to the presence of the army at Dalton Barracks. As the local area is relatively affluent and house prices are slightly lower outside the city boundaries there are high levels of commuting to and from Oxford, although house prices are still considerably higher than the national average, so unavailability and unaffordability of housing is a significant issue around Oxford as well as within the city. The rural settlements are also functionally dependent upon the market town of Abingdon, especially for facilities and services that have recently been lost from the parishes themselves, such as a GP surgery and a children's centre, and the reliance of local residents upon these larger settlements makes apparent deficiencies in the local transport infrastructure, such as a lack of pavements, cycle paths and streetlighting, narrow rural roads that cannot easily be adapted and limited bus services in addition to ongoing concerns regarding weight of traffic, noise and pollution, and speeding. The rural character and open views are highly valued by local residents, and there is a strong sense of community spirit, with a diverse array of social activities and interest groups taking place at the Community Centre in Wootton and the Church Hall in St Helen Without. The allocation of a Strategic Development Site to the two parishes covered by the WSHWNP, with its potential to establish a new settlement several times larger than the current largest settlement in the parishes clearly brings a host of challenges to those involved in the neighbourhood plan, which seeks to maximise potential benefits in terms of new facilities and minimise potentially detrimental impacts in terms of traffic, loss of Green Belt and the merging of settlements, that might result from the proposed development.

Wootton and St Helen Without Neighbourhood Plan

The Localism Act 2011 introduced a new statutory framework for neighbourhood planning, with the requirements placed on neighbourhood planning teams in terms of content, process and approval specified in the Neighbourhood Planning (General) Regulations 2012 (updated in 2018). Essentially, the core steps in the process are:

1. A qualifying body – usually a parish council or a community forum in unparished areas – decides that it wishes to produce a neighbourhood plan and establishes a steering group or community forum as a distinct body to undertake this work and engage with the community as broadly as possible.
2. The qualifying body proposes a Designated Area (the area to which the finished plan will apply) which is subject to consultation with surrounding parishes, interested parties and the community, and which must be approved by the local authority.
3. Local issues, concerns, aspirations and desires are identified through extensive and thorough consultation. Commonly, this involves public meetings and workshops, attendance at community events, online surveys, household questionnaires, and sometimes tailored events for target groups, such as young people or businesses.
4. Other evidence about the Designated Area is also obtained, through both existing sources such as that available through the local authority or interest groups, and new sources, including reports on specific issues – such as social inclusion or landscape character – commissioned by the steering group or community forum for the purposes of the neighbourhood plan.
5. Based upon this evidence, the steering group or community forum drafts a series of policies to address the issues that have been identified. Commonly, these address affordable housing, environmental measures, open green space, eco-design and infrastructure. They might also designate areas as Local Green Spaces to protect them from development if they can be demonstrated to be locally significant and valued, allocate additional sites for development if there is a need for a specific type of development, and designate Heritage Assets for protection.
6. A draft neighbourhood plan, called the Pre-Submission Consultation Draft, is then released for public consultation for at least six weeks and must be sent to a range of statutory bodies including the Environment Agency, Natural England, and Highways England. Subsequent to this consultation, the steering group or neighbourhood forum amends the draft plan in accordance with the consultation feedback as they consider appropriate and must compile a report outlining how they have responded to the consultation feedback.

7 The revised plan can then be submitted to the local authority, along with the required supporting documentation, which includes:

   a A Strategic Environmental Assessment, a Sustainability Appraisal and/or a Habitats Regulations Assessment, as appropriate, or a Screening Opinion to explain why such assessments are not required. In general, if a neighbourhood plan allocates additional sites for development or is likely to have significant environmental impacts beyond those incurred by the relevant local plan, some form of environmental appraisal will be required.
   b A Basic Conditions Statement, which confirms that the neighbourhood plan has been developed in accordance with legal requirements. These include conformity with the NPPF, the relevant local plan, and the environmental regulations above.
   c A Consultation Statement, which details the varied forms of consultation undertaken, the main outcomes of that consultation, and the account of how the steering group or neighbourhood forum has responded to the formal consultation on the Pre-Submission draft.

8 Once the local authority has satisfied itself that the neighbourhood plan is ready to be examined by an independent planning inspector, it arranges for this examination to take place, the purpose of which is to establish that the neighbourhood plan meets the Basic Conditions requirements.
9 Once the neighbourhood plan has been confirmed as meeting the Basic Conditions requirements by both the independent examination and the local authority, a referendum is then held in the Designated Area. If more than 50% of voters approve the plan, the neighbourhood plan is officially 'made' and at this time it becomes a formal part of the planning framework and is used by the local authority to help determine planning applications in the Designated Area.

This is a protracted and onerous process due primarily to the need for robust evidence to underpin the proposed policies, especially if Local Green Spaces or other designations are proposed, and due to the time requirements associated with specific steps in the process. For example, the statutory period for public consultation is six weeks, which applies to the Designated Area, the Pre-submission Consultation Draft and – where necessary – the Strategic Environmental Assessment. In addition, the later stages of the process add further delay, such as another six-week period for publicity prior to examination and arranging the independent examination and organising the referendum which have no specific time limits, so these latter stages could take months, especially as they are managed by the local authority, which might have other priorities or in extreme cases might be obstructive.

The two parish councils formally agreed to produce a joint neighbourhood plan and issued a call for volunteers in the community newsletter for each parish late in 2016. The steering group held its first meeting in March 2017, by which time the application for the Designated Area had already been submitted by the parish councils, and the Designated Area was confirmed in July 2017. Between July 2017 and June 2018, we reviewed the available evidence, commissioned our own Character Assessment and undertook an extensive consultation process. This included public workshops to identify local strengths, weakness, opportunities and threats; attendance at a local summer community festival; a questionnaire sent to every household in the Designated Area, including a tailored version for young people; pop-up events and door-knocking to encourage completion of the questionnaire; special events for young people and for those with additional needs (for example, mobility difficulties); an online community mapping tool to identify more and less desirable features of our area; a workshop tailored for local businesses; and a workshop designed to 'test the evidence' for policies for which existing evidence was not deemed sufficiently robust.

During this period, we also compiled the emerging outcomes of our work and drafted a set of policies and the main text for the plan itself. We engaged expert support to help us draft policies that would be appropriate for a neighbourhood plan (in other words, policies that are land use and planning related) and that would be appropriately worded for the determination of planning applications (in other words, policies that give clear guidance on how a planning application should be determined).

Pre-Submission consultation closed early in July 2018, and the rest of that month was spent determining how to respond to the comments received, amending the plan as appropriate, completing the consultation statement, finalising the Screening Opinion in relation to environmental regulations (as we did not allocate any additional development sites and anticipated no significant environmental impacts), commissioning the Basic Conditions Statement and doing the design work to transform the text-only draft into the final document for submission.

The WSHWNP was formally submitted to the VWHDC on 6th August 2018, along with the required supporting documentation, a map of the Designated Area, a copy of the analysis of the questionnaire data, and a copy of the Character Assessment that we had commissioned. At the time of writing this chapter (April 2019), we are partway through the examination process. Allowing time for the examination process to be completed, for the neighbourhood plan to be amended, and for the referendum to be organised and held, it is possible that our plan will not be made until September 2019, more than a year since we formally submitted it to the VWHDC.

Our 14 policies are organised under three headings (WSHWNPSG, 2018):

1 A *spatial strategy* for the Designated Area, to protect the Green Belt, the rural, open nature of our area and the separation of our individual settlements. The spatial strategy designates Local Green Spaces, Strategic Green Gaps between our settlements and Strategic vistas.
2 *Infrastructure needs*, addressing housing needs, the timing of infrastructure provision, mitigation of known or anticipated transport challenges and opportunities for enhancing transport, and community and business infrastructure.

3 A *design guide* to ensure that development 'fits in with' the rural and historic character of its surroundings, to encourage high standards of sustainable design, and to protect locally significant heritage assets.

However, in our case, the plan itself is only half the story, as the timing of the production of our neighbourhood plan relative to the timing of the production of LPP2 raised specific challenges but also potential benefits, as detailed in the next section.

Neighbourhood plan and local plan

A core requirement of neighbourhood planning is that the neighbourhood plan must be in broad conformity with the relevant local plan and cannot go against the strategic aims of that local plan. Importantly, though, it is the *adopted* local plan with which the neighbourhood plan must conform. In our case, the adopted local plan was the Part 1 plan, which proposed numerous smaller development sites rather than the strategic site at Dalton Barracks and Abingdon Airfield. At the same time, LPP2 was underway and was likely to be adopted in a form not dissimilar to the draft, and although not specified in the regulations, emerging local plans are normally granted weight in the examination of neighbourhood plans (Parker et al, 2019). We therefore needed to conform with Part 1 but be mindful of Part 2 as it would be the development proposals in Part 2 that would apply to the area for the period of both plans (up to 2031).

Significantly, the proposed allocation of a Strategic Development Site at Dalton Barracks and Abingdon Airfield was announced by VWHDC at a public meeting on the same day as the first formal meeting of the WSHWNP Steering Group: the neighbourhood plan volunteers went straight from the public meeting to their first steering group meeting. Unsurprisingly, the prospect of such a large development in the middle of our area was something of a rallying force, especially as in its initial iteration, it was proposed to merge the new development with two pre-existing settlements of Shippon and Whitecross. Shippon is a small historic village, which is named in the Domesday Book and retains its historic character at its core, with alms houses, village pump and former schoolhouse, and which – although adjacent to the MoD site – is distinct from that site. Whitecross is an even smaller linear village with a mix of land use including agricultural, residential and light industry, and which has open rural views to both sides. The residents of neither village wished to be merged with the new development and, as several of the volunteers on the steering group were residents of those villages, it would be fair to say that passions have run high throughout the neighbourhood planning process.

On the one hand, the neighbourhood plan would need to respond to this development proposal but could not refuse the allocation despite the strength of local feeling due to the requirement for neighbourhood plans to support the strategic objectives of the relevant local plan. On the other hand, the production of LPP2 would bring opportunities for local views to be voiced through the statutory consultation process for the emerging local plan while our own neighbourhood plan brought its own opportunities for consultation. We were in an unusual position, then, in that we were obliged to support the proposed allocation for the purposes of the neighbourhood plan but as representatives of our communities we also needed to convey the strong local opposition to such a large development being superimposed onto our small rural settlements. This developed into an attempt to influence the emerging LPP2 in such a way that it became more acceptable to local residents so that the need for the neighbourhood plan to conform with the final version of LPP2 would no longer be problematic. In essence, we developed a strategic approach to re-forging, as effectively as we could, the relationship between the two levels of plan: local and neighbourhood.

To this end, we focused considerable effort on influencing the emerging LPP2, including through participation in the public examination of that document. This consultation process was not an open affair, in which anybody could comment on anything they liked, but was highly structured and focused on whether the draft LPP2 was sound, legally compliant and consistent with the Duty to Cooperate. The emphasis, then, was technical rather than political. Somewhat counter-intuitively, although this emphasis on technicalities is symptomatic of a post-political drive to eradicate opportunities for political antagonism (Mouffe, 2005; Parker and Street, 2015; Etherington and Jones, 2017; Kenis, 2018), in practice it granted us greater freedom to input local concerns into this process than we perhaps anticipated as we could bring our political concerns to bear on the technical questions. However, this greater freedom was dependent upon our ability to articulate local concerns in terms of soundness and compliance. In November 2017, we submitted eight representations to the consultation for LPP2, covering issues such as the Green Belt, the separation and character of settlements and concerns over transport infrastructure.

When the revised version of LPP2 was released in February 2018, the residents of Whitecross were much relieved to see that their settlement was no longer to be removed from the Green Belt or to be merged with the new development but would retain green space between the two settlements, although a short distance at the top of Whitecross was to remain adjacent to the boundary of the new settlement. Residents of Shippon were less fortunate, as there was no change in the proposal to merge the new development with Shippon. At this stage, the evidence gathering, consultation and drafting work for the neighbourhood plan was well underway, and our policies were increasingly oriented towards a state of 'last resort' to do what we could through the neighbourhood plan if we were unsuccessful in influencing the local plan to the extent that we would like.

As part of that initial consultation activity, we indicated the desire of the WSH-WNPSG to attend the public examination of LPP2, at which an independent planning inspector tests the local plan for soundness, legal compliance and other requirements. In advance of that examination, a series of questions was released that related to the specific matters that would be addressed each day, to which those invited to speak at the examination could respond. We were invited to speak at the examination in relation to a matter entitled 'Dalton Barracks', and we provided responses to the questions released for that matter. As with the consultation, this stage in the process was highly structured, allowing input only in relation to those questions that had been released and allowing reinforcement but not repetition of feedback provided during the earlier round of consultation. Again, this worked relatively well for us as our consultation on the neighbourhood plan had progressed significantly since our previous representations and we were able to update those earlier contributions with the outcomes of our own consultation to demonstrate the strength of local feeling on the issues addressed as well as to reinforce our earlier comments.

The examination of LPP2 in July 2018 made apparent the unequal distribution of power and resources between different stakeholders as the LPA representatives, the representatives for the site promoter, and the developers in attendance were all paid professionals and were attending for the purpose of their work but the representatives of the parish councils and the neighbourhood plan did not have professional experience in planning and were attending on a voluntary basis, in some cases taking time out of their own work to do so. The LPA was also allowed to field several representatives, including a barrister, while other attendees were restricted to only one representative, and neither the parish councils nor the neighbourhood plan had sufficient resources to bring their own legal counsel. Despite these inequalities, the examination provided a further opportunity to ensure that local concerns were considered in the development of LPP2, and the WSHWNPSG and both parish councils ended their two days at the examination feeling that they had not been excessively marginalised from the hearing as they feared might be the case.

In terms of outcomes, these are not yet known. The LPA was asked to consider certain concerns that we had raised, for example about the need for separation of settlements and mitigation of light pollution, and we were pleased to be able to provide the inspector with a copy of our neighbourhood plan following our participation in the examination of LPP2, especially as it had not yet been made. This is important as the closer a neighbourhood plan is to completion, the greater the weight that it carries in processes such as the public examination of LPP2. Consequently, the relative timing of the production of these two documents became critical and is addressed in more detail in Chapter 4. Our ability to participate in the public examination of LPP2 from such an advanced position should increase the chances of securing community objectives, but we are still a long way from our neighbourhood plan being made and being reasonably satisfied with the event is no guarantee of the outcome.

Conclusion

This chapter has set the scene for the work that follows, situating the specifics of our own experience in the context of the local circumstances that feed into neighbourhood planning priorities and preferences, opportunities and constraints, and of the spatial configuration of local government structures and relations in the surrounding area. These governance arrangements, and the ways in which they have the potential in the aftermath of the Localism Act 2011 to impact upon decision-making with respect to economic growth and planning, have also been summarised as a means of signposting some of the broader influences on neighbourhood planning activities. This chapter, then, weaves a narrative thread from the historical context for the emergence of this newest form of localism, through the shifting geographies of local government and local governance, to the specific intersection of localist agendas, governance arrangements and neighbourhood circumstances that has spawned and continues to inform the WSHWNP.

References

Bradley, Q (2018) Neighbourhood planning and the production of spatial knowledge. *The Town Planning Review* 89 (1): 23–42

Bradley, Q and Brownill, S (2017) Reflections of neighbourhood planning: towards a progressive localism. In S Brownill and Q Bradley (eds) *Localism and neighbourhood planning: power to the people?* Policy Press: Bristol, p251–267

Bradley, Q, Burnett, A and Sparling, W (2017) Neighbourhood planning and the spatial practices of localism. In S Brownill and Q Bradley (eds) *Localism and neighbourhood planning: power to the people?* Policy Press: Bristol, p57–74

Brownill, S (2017) Neighbourhood planning and the purposes and practices of localism. In S Brownill and Q Bradley (eds) *Localism and neighbourhood planning: power to the people?* Policy Press: Bristol, p19–38

Brownill, S and Bradley, Q (2017) Introduction. In S Brownill and Q Bradley (eds) *Localism and neighbourhood planning: power to the people?* Policy Press: Bristol, p1–15

Cabinet Office (2015) *Deregulation act* www.legislation.gov.uk/ukpga/2015/20/contents/enacted

Colomb, C (2017) Participation and conflict in the formation of neighbourhood areas and forums in 'super-diverse' cities. In S Brownill and Q Bradley (eds) *Localism and neighbourhood planning: power to the people?* Policy Press: Bristol, p127–144

Davoudi, S and Madanipour, A (2015a) Localism and the 'Post-social' governmentality. In S Davoudi and A Madanipour (eds) *Reconsidering localism*. Routledge: New York and London, chapter 5. (no page numbers) Accessed 6 Dec 2018

Davoudi, S and Madanipour, A (2015b) Introduction. In S Davoudi and A Madanipour (eds) *Reconsidering localism*. Routledge: New York and London, chapter 1. (no page numbers) Accessed 6 Dec 2018

DCLG (2011) *The localism act* www.legislation.gov.uk/ukpga/2011/20/contents/enacted

DCLG (2012) *Neighbourhood planning (general) regulations* www.legislation.gov.uk/uksi/2012/637/contents/made

DCLG (2012/2018) *National planning policy framework* www.gov.uk/government/publications/national-planning-policy-framework-2

DCLG (2016a) *Cities and local government devolution act 2016* www.legislation.gov.uk/ukpga/2016/1/notes/division/1/index.htm

DCLG (2016b) *Locally led garden villages, towns and cities* www.gov.uk/government/publications/locally-led-garden-villages-towns-and-cities

Etherington, D and Jones, M (2017) Re-stating the post-political: depoliticization, social inequalities, and city-region growth. *Environment and Planning A: Economy and Space* 50 (1): 51–72

Gallent, N and Robinson, S (2013) *Neighbourhood planning: communities, networks and governance*. Policy Press: Bristol

Kenis, A (2018) Post-politics contested: why multiple voices on climate change do not equal politicisation. *Environment and Planning C: Politics and Space* DOI:10.1177/0263774X18807209

Milligan, C and Conradson, D (2006) Contemporary landscapes of welfare: the 'voluntary turn'? In C Milligan and D Conradson (eds) *Landscapes of voluntarism: new spaces of health, welfare and governance*. Chapter 1 (no page numbers) Accessed 14 Jan 2019

MoD (2016) *A better defence estate* www.gov.uk/government/publications/better-defence-estate-strategy

Mouffe, C (2005) *On the political*. Routledge: Abingdon

Owen, S., Moseley, M and Courtney, P (2007) Bridging the gap: an attempt to reconcile strategic planning and very local community-based planning in rural England. *Local Government Studies* 33 (1): 49–76

Parker, G (2012) *Neighbourhood planning: precursors, lessons and prospects*. 40th Joint Planning Law Conference, Oxford www.quadrilect.com/Gavin%20Parker.pdf Accessed 15 Dec 2018

Parker, G (2014) Engaging neighbourhoods: experiences of transactive planning with communities in England. In N Gallent and D Ciaffi (eds) *Community action and planning: contexts, drivers and outcomes*. Policy Press: Bristol, p177–200

Parker, G (2017) The uneven geographies of neighbourhood planning in England. In S Brownill and Q Bradley (eds) *Localism and neighbourhood planning: power to the people?* Policy Press: Bristol, p75–91

Parker, G, Salter, K and Wargent, M (2019) *Neighbourhood planning in practice*. Lund Humphries: London

Parker, G and Street, E (2015) Planning at the neighbourhood scale: localism, dialogical politics, and the modulation of community action. *Environment and Planning C: Government and Policy* 33: 794–810

Vigar, G, Gunna, S and Brookes, E (2017) Governing our neighbours: participation and conflict in neighbourhood planning. *The Town Planning Review* 88 (4): 423–442

VWHDC (2016) *Vale of white horse district council local plan 2031, part 1: strategic sites and policies* www.whitehorsedc.gov.uk/services-and-advice/planning-and-building/planning-policy/new-local-plan-2031-part-1-strategic-sites

VWHDC (2018) *Vale of white horse district council local plan 2031, part 2: detailed policies and additional sites* www.whitehorsedc.gov.uk/services-and-advice/planning-and-building/planning-policy/local-plan-2031-part-2

Wargent, M and Parker, G (2018) Re-imagining neighbourhood governance: the future of neighbourhood planning in England. *The Town Planning Review* 89 (4): 379–402

Wills, J (2016a) *Locating localism: statecraft, citizenship and democracy*. Policy Press: Bristol

Wills, J (2016b) Emerging geographies of English localism: the case of neighbourhood planning. *Political Geography* 53: 43–53

WSHWNPSG (2018) *Wootton and St Helen Without Neighbourhood Plan 2018–2031* www.whitehorsedc.gov.uk/sites/default/files/WSHSNP%20Plan.pdf

2 Conceptual Constructions of Place

Introduction

If planning as a field of practice is about the relationship between people and place (Beauregard, 2013), then neighbourhood planning is about the relationship between a certain sub-group of people (a community) and a specifically local place (their neighbourhood), but within the context of the broader space of planning as a field of practice. My focus in this chapter – as a geographer – is on the nature and significance of place in neighbourhood planning, and more specifically, on the different ways in which we can think about matters of place and space. The discussion is contextualised within a general interest in the structures of local government and planning but is informed by specific investigations of the nature and significance of place in the existing literature on neighbourhood planning.

While neighbourhood planning seeks to incorporate a more local form of place knowledge into the planning of space and the construction of new places in a purported multiplication and democratisation of planning (Bradley, 2018; Lennon and Moore, 2018), I suggest in this chapter that several shortcomings persist, including: 1) the prevailing dominance of binary thinking (e.g. space versus place, lived versus abstract), which highlights the inherent limitations of incorporating such local knowledge into formal planning processes and documents; 2) an excessive focus on the positivity of place attachment and the making of place at the expense of considering the detrimental implications for place attachment through the unmaking of place as part of the planning project, and 3) the excessive emphasis on the outcomes of neighbourhood planning as opposed to the process, practice and experience of its unfolding, whereby closer examination of specific instances of neighbourhood planning can help to unpick these varied and interwoven considerations. This chapter therefore lays the conceptual groundwork for the close examination delivered in the subsequent three chapters, each of which addresses a different perspective on this more active and interactive consideration of the construction of place in neighbourhood planning: thematic, strategic and performative.

Dimensional and networked understandings of space

One of the most commonly reported features – and criticisms – of neighbourhood planning is the extent to which a neighbourhood plan is dependent upon and constrained by higher level plans and policies (Gallent and Robinson, 2013; Wills, 2016; Bradley and Brownill, 2017; Brownill and Bradley, 2017). While at present there is no regional layer in the planning system, there remains a clear hierarchy of planning policy, with the National Planning Policy Framework (NPPF) at the top setting the broad parameters for all planning in England, the local plans of Local Planning Authorities that set out the strategic management of development in an authority's area in the middle, and – for those areas with a neighbourhood plan – neighbourhood plans at the bottom, which specify more detailed policies on non-strategic issues of local concern. Within this structure, neighbourhood plans must conform with the relevant local plan and with the NPPF, while local plans must conform with the NPPF (DCLG, 2012).

We can, though, extend this vertical hierarchy further, as the NPPF embeds in the planning system a host of international requirements, for example in relation to human rights and environmental protection. This vertical arrangement, then, both mainstreams and enforces international obligations through the planning system: the public examination of both neighbourhood and local plans addresses – among other things – compliance with these international obligations as well as compliance with the NPPF, and compliance with those obligations through compliance with the NPPF.

Within this vertical structure, the localism agenda seeks to empower communities by installing neighbourhood planning as a new, lowest level in the planning system. By granting these plans legal weight once they are made (adopted), the local concerns and priorities enshrined within them become a formal part of the planning framework and inform the determination of planning applications that affect the area covered by the plan. An optimistic view of this addition to the planning framework perceives neighbourhood planning as disrupting the traditional top-down driving of planning in a significant reconfiguration of power relations and transfer of power to the people (Wills, 2016; Bradley and Brownill, 2017), but more pessimistic (realistic?) views see neighbourhood planning as enforcing compliance, reproducing power dynamics and reducing independence even if there is some slight enhancement in power at the neighbourhood level (Gallent and Robinson, 2013; Bradley, 2017a; Bradley and Brownill, 2017).

While there are some connections between the different levels of the planning system through the requirement for conformity, the different layers or levels are primarily distinct from one another, with clearly defined purposes and content. Comparing local plans and neighbourhood plans illustrates this clearly as a local plan is deemed necessary to set the strategic direction for development in its area and a neighbourhood plan should only address non-strategic issues (Parker, 2012). The direction of influence is also assumed to be only one way: a neighbourhood plan or the team behind it are not considered to have any role or authority in influencing the local plan with which it must conform. It should simply 'toe the line' once the local plan has been adopted, although there are a few alternative views on this matter (Parker and Salter, 2017). In essence, then, the planning framework is a highly containerised spatial imaginary in which different roles, responsibilities and powers are allocated to different levels, which have strictly prescribed influential connections between them.

While the planning system is conceived in vertical terms, the space of local governance is often conceived horizontally, as a patchwork of spaces over which different civic bodies have responsibility. The system of parish councils is a clear example, but so too is the distribution of district or county councils. Whether at the parish, district or county level, different councils have the same responsibilities, powers and duties but will have authority over a different geographical area. Again, this is a highly containerised spatial imaginary as although different councils will work with each other on cross-boundary issues such as transport or to achieve cost savings, for example through the delivery of shared services, they only have autonomy over their own local authority area. Similarly, local people can only be resident in one or another parish or district council area, even if they might have a second home elsewhere, they might work elsewhere, or they might have other interests elsewhere.

Immediately, though, we need to complicate this simplistic horizontal spatiality, as – on the one hand – there are multiple horizontals, and – on the other hand – the horizontals are not pure or straightforward. In parished areas, for example, parish councils typically sit below district councils, which sit below county councils, so there is a vertical organisation to the horizontal distribution of local governance. At the same time, the horizontals can be both 'holey' and 'lumpy'. They can be holey in the sense that – for example – not everywhere is covered by a parish so there are gaps in that level of governance. They can be lumpy in the sense that to be horizontal does not mean to be flat, as some forms of local government – unitary authorities – have the duties, responsibilities and powers of both a district and a county council.

Evidently, crude distinctions between vertical and horizontal conceptions of space in planning and local government are overly simplistic and misleading (Brownill, 2017a) as the structures of governance and the planning system that operates through them have evolved over generations and centuries, partly in an organic fashion and partly due to intentional design. Neither the vertical nor the horizontal themselves is straightforward, nor do they map neatly onto each other. However, it is also worth noting that these structures are not static but are liable to change. Before regional spatial planning was abolished in 2004 (Gallent and Robinson, 2013), planning in Oxfordshire would have been informed by organisations such as the Government Office for the South East, but this level of governance has now been removed. In addition, governance reviews are carried out periodically and can lead to the redrawing of boundaries, or to the addition, deletion or merging of areas of governance. The localism agenda itself is a prime example of this malleability and instability, introducing a new layer to the planning hierarchy in the form of neighbourhood planning and inviting applications for reconfigured governance arrangements at district and county level. Consequently, to understand the spaces of planning and governance, and the relation between them, we need a more comprehensive, refined and more active conceptual tool.

Networked understandings of space move us one step closer to this, as they do not assume orderly, static or simplistic spatial structures, but instead emphasise multiplicity, contingency and messiness. Rather than a neatly nested hierarchy of spaces of governance through which the planning system runs as singular stream of influence, any one space of governance can be multiply and messily connected to numerous others at a host of levels and not in a manner that is constrained by proximity or adjacency.

On such an understanding, planning influence does not pass from layer A to layer B to layer C and so on, but can jump from layer A to layer C, or further. In the case of neighbourhood planning, a neighbourhood plan must conform with both the local plan and with the NPPF, so that in the unlikely event that a local plan does not say anything about an aspect of planning in the NPPF that a neighbourhood plan mentions in relation to a local concern, the neighbourhood plan must conform with the NPPF independently of any conformity with the local plan. Similarly, councils working with other councils do not have to work just with those with which they share a boundary or only those at the same level as them, as strategic issues might span much further afield and involve multiple levels of governance. Major transport infrastructure projects – such as HS2 or the Oxford-Cambridge Expressway – are good examples here, as authorities remote from each other are affected in similar ways, while in any one area, multiple layers of governance will have their own perspectives.

Another benefit of thinking about space in terms of networks is that it broadens the range of actors and influences that is considered. Thinking about planning as vertical and governance as horizontal restricts our thinking to the formal structures of planning and governance, but neither planning nor governance is confined to these machinic structures. Even before the formal publication of The Localism Act 2011, local government was more than just 'the council' as the neoliberalisation of the Thatcher years invited the private sector into council operations and broadened the practice of councils from government to governance, while this engagement was broadened further to other stakeholders including other public bodies, the third sector and the public in efforts to coordinate services through Local Strategic Partnerships (Gallent and Robinson, 2013; Brownill, 2017b). The localism agenda reinforces this trend and formalises it in the context of planning, by involving non-professionals in its system and practice and affording the plans that they generate legal weight.

By way of a neighbourhood planning example, the enforcement of obligations under international environmental regulations is only partly conducted through the formal examination of the plan and the required conformity with the NPPF, as the requirement to consult with statutory bodies – for example, the Environment Agency – provides another check on this aspect of a neighbourhood plan, while local interest groups either professional or voluntary can also involve themselves in this consultation process. In this example, the management of the local environment is influenced by international law, national planning structures, statutory bodies that might themselves be located remotely from the local area, and local non-professionals. The local here is not just local and is not just jumping scale

(Brownill, 2017a), but the local is at the same time both local and global as one of the factors that defines local character is the international legislation governing designated sites for habitats and wildlife. At the same time, it is not just local and global, as the statutory bodies operate nationally, leading to the collapse of any scalar distinctions that we might recognise in everyday speech as each scale is implicated in and constituted by the others (Agnew, 2005; Massey, 1993, 2005). Space in this understanding is far more than a background or container for action but is a medium through which the character of place is reproduced and challenged in multidirectional and multifaceted ways (Cresswell, 2014). Not only are dualisms such as local and global dubious in a conceptual sense (Agnew, 2005; Dorn et al, 2010; Cresswell, 2014) but they are unsustainable in an experiential sense as those involved in neighbourhood planning are at once concerned with the neighbourhood, engaged with the local, constrained by the national and informed by the global, with each of these influences feeding through an array of media, whether formal, statutory, informal, voluntary, professional, thematic, economic or interest driven.

At its heart, then, the space of neighbourhood planning can be considered to be vertical, in that the content and requirements of neighbourhood plans are in no small way specified by higher levels in the NPPF; horizontal in the sense that Designated Areas are spaces that are bordered by other parishes, districts, Designated Areas, etc; and networked in that not only do the horizontal and vertical interact with each other but in that a whole host of other actors, agencies and organisations are enrolled in the neighbourhood planning process. An illustrative example might be the formalisation of our Designated Area, which covers the vast majority but not all of the two parishes of Wootton and St Helen Without. It evidences the verticality of the planning system in that the procedure for determining a Designated Area is specified in national regulations and the proposed Designated Area must be approved by the relevant local authority. It evidences the horizontality of local government in that neighbouring parish and town councils responded to the consultation, which led to the exclusion of one small part of St Helen Without from our Designated Area due to its anticipated functional dependency on Abingdon rather than Shippon in light of proposed housing development on the site concerned. It also evidences networked space in that the MoD and MPs (formally part of neither the planning nor local government systems) were engaged in seeking to include Dalton Barracks and Abingdon Airfield within the Designated Area given the anticipated disposal of the site by the MoD. Consequently, neighbourhood planning is simultaneously all three spatial imaginaries: it is bound to the vertical, contested by the horizontal, and driven by the actors and dynamics in the network. At the same time, neighbourhood planning is bound to the horizontal as it is attributed to one specific layer of planning and governance; it is contested by the vertical as both individual residents from below and both district and county councils from above can comment critically on the Pre-Submission draft; and it is also obstructed by the network through the clash and negotiation of competing interests. Neighbourhood planning, then, is never just vertical or horizontal, although it is both, nor is it ever just local as the local, national and global intermesh with, mediate and constitute each other in an interconnected network.

This is a more progressive sense of space that seeks to capture the throwntogetherness of social and spatial relations (Massey, 2005) where connections are extensible, unstable and generative, but this is not to suggest that this throwntogetherness occurs in a vacuum as the resulting spatialities are not entirely footloose. While there is an indeterminacy and open-endedness to this understanding of space, there is also a contingency as the extensions and instabilities occur against a backdrop of that which went before. Similarly, opening the doors of the planning framework to involvement by local people does not generate frictionless space in which all people and communities are equal in their ability to engage with neighbourhood planning or in the capacities of the relevant local authority to listen to, engage with and respond to the needs of the neighbourhood planning team and the communities that they represent. To that extent, there is a certain path dependency within neighbourhood planning, whereby local social and spatial divisions can be reinforced as those people and areas with greater capacity, financial resources or time will be better placed to engage with neighbourhood planning than those lacking such necessities (Gallent and Robinson, 2013; Wills, 2016; Bradley, 2017a; Bradley and Brownill, 2017; Brownill, 2017b; Colomb, 2017; Parker et al, 2017; Vigar et al, 2017; Wargent and Parker, 2018).

Thus, although this networked approach to planning – bringing multiple interested parties with different perspectives together to develop planning policies – aims to reduce conflict and speed up development by reducing the number of appeals against both draft planning documents and planning applications (Gallent and Robinson, 2013; Bradley et al, 2017) – these outcomes are by no means certain or necessarily co-deliverable. The evidence in this regard appears mixed, as although in 2015 housebuilding rose by 10% and neighbourhood plans allocated more sites for development than local plans, suggesting success in terms of housing output, community-developer antagonism increased due to perceptions on the part of developers that neighbourhood plans are obstructing speculative housebuilding (Bradley et al, 2017). Bringing different perspectives together can inflame rather than reduce conflict between parties, especially if communities seek to introduce or encourage an alternative model of development that favours small scale development with the explicit intention of meeting local need, and if pre-existing imbalances in power between the parties are re-inscribed rather than reconfigured (Gallent and Robinson, 2013, Bradley, 2017a; Bradley and Brownill, 2017; Bradley et al, 2017).

The preceding discussion has progressed from basic understandings of the space of planning as vertical and the space of governance as horizontal through a networked understanding of how these work together in a non-straightforward fashion, featuring myriad actors, influences and inter-scalar connections that inform the space of neighbourhood planning, as shown diagrammatically in Figure 2.1.

However, despite recent recognition of the need to consider how these assemblages of actors come together and play out in place (Brownill, 2017a), it is not just how these actors come together that is of concern but also the extent to which

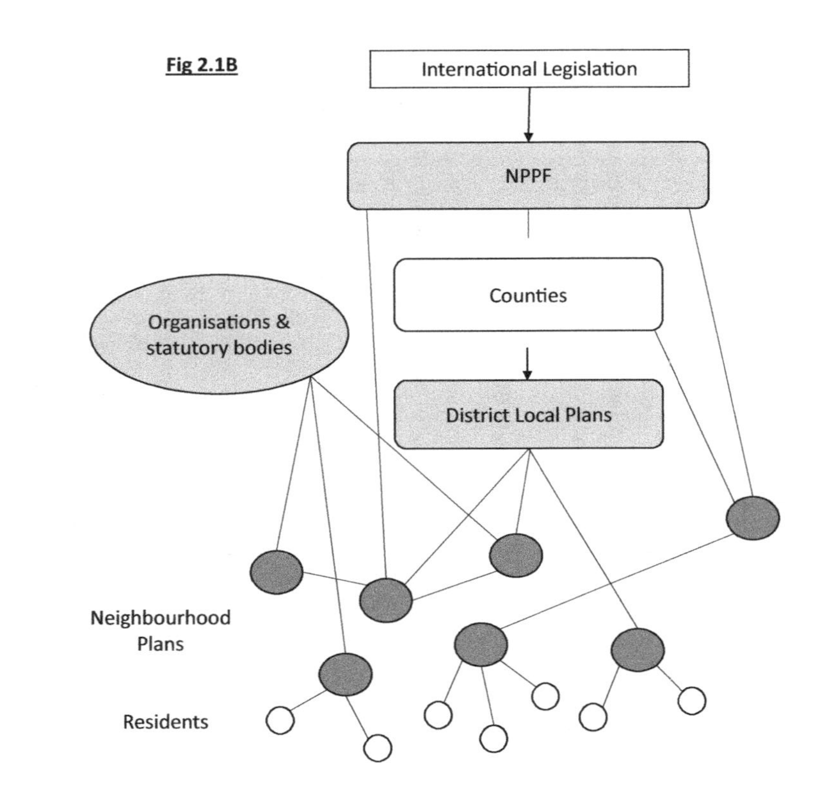

Figure 2.1 Spatial imaginaries of neighbourhood planning

Source: Author

and ways in which actors continue to be held apart within the assemblage, the potential within the network for blockages and tensions to be resolved, and how place is subsequently constructed, deconstructed and reconstructed. Such concerns suggest that the making of place through neighbourhood planning is not necessarily a cosy coming-together, nor is it necessarily a unidirectional process of construction. The opportunities for performativity on the part of the diverse actors involved are many, varied and unequal, and these relative capacities to drive the network lie at the very heart of the effectiveness of neighbourhood planning.

Performative and affective understandings of place

While a networked understanding of the space of neighbourhood planning can help us to identify the number and range of actors and influences involved, it still only takes us partway to our destination. It can help us to understand the different actors and levels that come together in fuzzy power relations that are not solely scalar (Brownill, 2017a) but I would question whether it helps us to understand *how* they come together and play out in the process of constructing place. While it is valuable to consider how interests reach into and out of the neighbourhood, the power to assemble the space of the neighbourhood around specific objectives, and how assemblages vary across both time and space (Brownill, 2017a), my main concern is with the assumption that this coming together and playing out takes place *in the same place*. My interest is in how place itself is variously constructed by different parties within a single neighbourhood plan process, and in how the same place can be variously and repeatedly made and unmade during the course of neighbourhood planning. To this extent, we need to think about place as multiple, performed, contested and unstable as well as complexly interacted and thrown together (Massey, 2005), and about the implications of this for the effectiveness of a neighbourhood plan, given the power dynamics identified as being at play and in the very nature of planning. In this section, I approach these matters in a conceptual fashion, exploring first the performativity of place as exceeding and eliding the determinisms of planning, then considering the affectivity of place attachment escaping capture in the formal language of planning, and finally interrogating the nature of and implications for place attachment in neighbourhood planning.

In the first instance, we can consider the different understandings of place already acknowledged in the literature on neighbourhood planning, which – as in geography – tend to revolve around binaries. Place is one of the most important terms in geography and can be considered a way of understanding the world (Cresswell, 2014). Place is commonly contrasted with space in that space is seen as an abstract concept, as in the non-specific ideas of horizontal, vertical and networked space, but place is effectively space with meaning, such as home, country or a favourite chair. Space is invested with meaning in both individualised and socialised ways, such that there are both common frameworks of spatial meanings that relate to certain types of place (e.g. a hospital, school or prison) and idiosyncratic meanings attached to specific places based on our own personal experiences of such places. Immediately, then, we have a multiple understanding of place in which the varying positions and experiences of individuals affords them different understandings of the world despite being in the same geographical space.

On the one hand, this is significant to neighbourhood planning because it tends to reinforce social and spatial divisions, so if only certain sections of a community are involved in neighbourhood planning it will be primarily their understanding of place that will be written into the plan. On the other hand, it is also significant to neighbourhood planning because it raises the issue of the different communities of practice that are involved in that process: professional experts on one side and non-professional residents on the other; those promoting development on one side and those seeking to control it on the other. Not only do they have different agendas, but they also have fundamentally different ways of understanding place, which influences how they engage with the neighbourhood planning process. Just as the understandings of place that underpin their engagement with neighbourhood planning are contested, so too is the neighbourhood plan that is generated through their engagement.

For instance, dualistic distinctions have been drawn in literature on planning between abstract, system or economic space on the one hand and lived or life space on the other (Fenster and Misgav, 2014; van der Pennen and Schreuders, 2014; Bradley, 2018); between the resident as an insider as opposed to the planner or developer as an outsider (Stephenson, 2010); between professional and everyday or community understandings of place (Stephenson, 2010; Fenster and Misgav, 2014; Bradley, 2018); between surface and deep or embedded engagements with place (Stephenson, 2010); and between place and site in planning circles (Beauregard, 2013). We also see the vilification of local and lay place knowledge as expressing nothing but NIMBYist resistance to development that any 'right-minded' person would consider to be in the public good, such that residents are not perceived to have anything legitimate to contribute to planning. Equally, those promoting community agendas consider the technocratic expertise of planning and the associated need to translate their aspirations into the language of planning to strip away the essence of local character and sense of place from their neighbourhood plan (Parker, 2012; Parker et al, 2015; Wargent and Parker, 2018). In other words, the local lived realm of the neighbourhood is often pitted against the rational abstract realm of planning, and the two are considered incommensurable.

In a crude sense, then, we can characterise planners and developers as working with space while local people live in place. Academically, it is acknowledged that both space and place – the abstract and the experiential – are required for spatial theory to be satisfactory (Cresswell, 2014) and neighbourhood planning can be seen as an attempt to integrate the local experience of place into the planning of space, but immediately we hit an obstacle. There is an inherent contradiction between the perpetual performance of the indeterminacy of place in the lived experience of place, reproducing place anew each day but doing so slightly differently within a contingent multiplicity, and the purpose of planning to make determinate that indeterminacy, channelling the contingent multiplicity into a planned and ordered singularity. Plans in formal planning documents might clearly demarcate areas that are shaded to indicate their primary or socially sanctioned purpose but in reality it is likely that the boundaries are not upheld so rigidly and their actual purpose might be very different to that portrayed on the plan (think, for example, of vacant retail premises used for residential purposes by squatters). Equally, while residential and retail zones and traffic systems for a new settlement might be laid out to channel traffic in certain directions and along certain routes, the detailed interconnections between future residents and facilities cannot be foretold in advance, such that human creativity will likely forge new paths or channel traffic flows in unexpected directions and with unanticipated effects. The daily performativity of place will only ever partially respect the determinisms of planning, especially in situations where whole new settlements are proposed as there is no local lived or practical experience of *that* place to incorporate into the planning of it.

The significance of practical constructions of place becomes especially clear when considered in contrast to more remote constructions of place which are often visually based as depicted on maps. For example, the assumption enshrined in LPP2 that the new development should be merged with Shippon seemingly on the basis that the village and the current barracks are proximate to each other on a map arguably neglects the clear distinctions between the two places at ground level. Building densities, architectural style, scale of buildings, and boundary treatments, not to mention the military fencing topped with razor wire and the military signage and security infrastructure, distinguish very clearly between the village and the barracks, but this only becomes evident through practical involvement in that place. Whereas in planning terms it is a simple if not obvious step to merge the new settlement at the barracks with the historic village of Shippon; for residents of Shippon, however, nothing could be less obvious because the practice of living in that place reveals and reproduces the differences between the two.

The idea of practical constructions of place links to the notion of a personal attachment or sense of place (Relph, 1976; Tuan, 1974, 1975; Norberg-Schultz, 1980; Bradley, 2017b, 2017c) generated through active engagement with place, but practical constructions of place are not necessarily associated with place attachment. What I mean by a practical construction of place, then, draws a distinction between Relph's authentic geography as an experience of place resulting from insider involvement on a human level (Relph, 1976) and Tuan's topophilia as a strong affectionate attachment to place generated through deep immersion in it (Tuan, 1974, 1975). A practical construction of place is simply the understanding of a place that comes from practical involvement with that place, which might or might not be associated with a strong affective or emotional attachment to that place.

This emotional bond with place is also performative, being generated through active engagement in those places (Tuan, 1974, 1975; Bradley, 2017b), and has been identified as being as important to wellbeing as interpersonal relationships. With emplacement understood as a condition of being in the same way that embodiment is (Bradley, 2017b), it is perhaps not surprising that wellbeing is so closely tied to place identity, or that experiences of sostalgia – a deep feeling of loss – can arise in the face of degradation of a familiar or cherished place (Jack, 2012; Cleary et al, 2017). Indeed, such degradation need not materialise in physical form but can take on affective significance as perceived denigration of the person even before the degradation of the place occurs. By way of example, the merging of Whitecross with the new settlement was at one point described as turning Whitecross into a more meaningful settlement pattern, while during the examination of LPP2 the proposed creation of a country park within the Strategic Development Site was described as transforming the landscape into something more interesting than its current open grassland. As a resident, I must admit I found such assertions objectionably arrogant: it is – at least in part – the linearity of Whitecross that gives the settlement its meaning, and the open grassland is a significant part of what makes our neighbourhood what it is. For residents, whose personal identity might well be bound up with the place of their locality, any suggestion that their settlement lacks meaning or their locality lacks interest might be taken to suggest that so do they. The meaning and interest of the locality for those who live there spring simply and precisely from the fact that their locality is what it is. Bound into the practical understanding of that locality, meaning and interest can be positive or negative, but where positive, perceived slurs against the place can be perceived as personal slurs, with their associated emotional responses.

While we can think of practical place knowledge and affective place attachment as distinct concepts, they are often deemed to coincide in practice. Indeed, the very intention of neighbourhood planning is to draw on the place knowledge and attachment of local residents in planning the future development of their area (Bradley, 2017b, 2017c; Vigar et al, 2017; Lennon and Moore, 2018). However, this constellation of local place knowledge and affective place attachment is simultaneously deemed to be a valuable contribution to planning by those advocating neighbourhood planning, and an irrational, problematic obstacle to sensible and much-needed development by developers, so simply incorporating place knowledge and attachment into the formalities of planning does not resolve the tension between the different understandings of place identified earlier.

Further, irrespective of the intentions and aspirations of the Localism Act, incorporating this local place knowledge and affective place attachment into the formal, technical language of planning is no easy task. Setting aside debates in non-representational geography about the potential or otherwise to articulate from our pre-reflective or embodied experience (Banfield, 2016), even if we could do so in everyday language, planning does not operate in everyday language but according to its own conventions. The felt experience of place that comes from phenomenological immersion in that place is simply not what planning is about, any more than the indeterminacy of the practical performance of place. The implications of this for neighbourhood planning are significant, not only because it suggests a potentially insurmountable chasm between the understandings of place on the part of immersed locals and abstract planners, but because it suggests that collaborative approaches are no better suited to resolving this conflict than previous approaches to planning if they cannot find an effective means of bridging this communicative and existential gap. Both the indeterminate performativity of place and the implicit, affective sense of place that arises from that seemingly render infeasible the aims of localism to incorporate local knowledge and sense of place into the formal planning of space. The simple fact that residents and communities might be motivated to engage with their sense of place and to seek to protect it for the future is no guarantee that their place attachment can be either *written into* their neighbourhood plan or effectively *protected through* their neighbourhood plan. Planning is about locations, distributions, connections and patterns, but it is not about the felt experience and significance of living with one's feet firmly on the ground in that place. One fundamental difficulty with neighbourhood planning, then, is the challenge of conveying the existential meaning and indeterminacy of place in a document and system that explicitly exclude such meanings from consideration.

A related issue, but one that is not thoroughly explored in the literature, concerns normative assumptions about place attachment. As noted earlier, active immersion in a locality such as a neighbourhood can generate a deep and powerful sense of belonging to that place and it is this strength of feeling that is being used to influence and govern behaviour through localism. It can certainly be a powerful motivator to get people involved in the first place and to sustain their involvement in a prolonged, arduous and aggravating process, but the literature has a tendency to treat place attachment as a bottomless pit of positivity and enthusiasm; a resource to be utilised (exploited?) for the purposes of planning, and specifically for the purposes of removing opposition to development. Attention to place attachment is growing within the neighbourhood planning literature, and it has recently been proposed that place identity and attachment can be enhanced through the neighbourhood planning process (Bradley, 2017b, 2017c). However, such affirmative views of place attachment in the context of neighbourhood planning seemingly neglect the potential for the neighbourhood planning process to damage place identity and attachment, or more specifically, fail to consider *how* that might happen. While the perceived threat to local and individual identity as well as place attachment posed by large-scale development has been acknowledged in the literature (Bradley, 2017b, 2017c), the emphasis in these instances is on the development and the outcome of the process. What, though, happens to place identity and/or place attachment along the way?

I would argue that place attachment and/or sense of place have considerable capacity to be unsettled in a whole slew of ways. First, the multiple and diverse understandings of the same place held by different actors in the neighbourhood planning process become clear, removing from residents any perception that they are the only ones who are involved with their own neighbourhood. This might not undermine their sense of unique familiarity, but it does open the door to alternative knowledges in a mirror-image manoeuvre of the opening of the planning system to lay and local knowledges. Secondly, different spatialities and temporalities come into play simultaneously. On the one hand a neighbourhood plan is focused on its own locality, but it also has to contribute to the objectives of the LPA for a much bigger area and respond to comments and concerns raised by actors at other spatial scales too; perhaps at the county or national level. On the other hand, the temporal framings used by different actors also vary, as local residents likely start from the 'here and now' and work forwards in time whereas LPAs likely start from the future 'then and there' and work backwards towards the present – especially in the context of large-scale, strategic sites – as their planning documents become more detailed and finalised over the shorter-term. Thirdly, as the process unfolds, the power dynamics become more and more apparent, as on the one hand other interested parties provide greater input through consultation and examination, and on the other hand the LPA becomes more heavily involved towards the end of the process. Finally, not only are the ultimate outcomes unknown but for any team of volunteers experiencing the process for the first time, even the rules of engagement are unknown in anything other than the crudest of terms, so the alternative possible outcomes cannot be reliably scoped. Importantly, all this is occurring while residents are still performing the indeterminacy of place in their daily immersion in their neighbourhood, but which place are they now performing and how consistent is this from one day to the next? The possibilities for performing place anew each day are multiplied and the degree of difference is exaggerated by the intervention of new actors in the place of the neighbourhood. Place attachment might become more entrenched one day as motivation to persist with the process in the hope that neighbourhood ambitions will be realised, but on another day place attachment might dwindle if the anticipation is that those aspirations will not be realised.

Place attachment, then, like place, is anything but uniform, stable or static and while it can be a resource it is also fragile and can be reduced or destroyed, with implications for the longevity of neighbourhood planning as a localist project. While it is possible for place attachment to be enhanced through neighbourhood planning, it is also possible that the opposite might occur, and that even if the outcome is heightened place attachment, the intervening experience might be very rocky indeed. It would be simultaneously ironic and regrettable if the very place attachment upon which neighbourhood planning is predicated was destroyed by the very process that claims to grant it legitimacy and power, but I fear that this is a very real possibility.

Alternatively, if we draw on non-representational thinking in geography to consider the practice of planning as a space of experience, we might forge a more optimistic view about the perceived disparities between the spatial experiences of residents and planners. With its emphasis on active experience through practical doings, non-representational geography is concerned less with formal representations that we passively observe and more with the performative and affective aspects of bringing those representations into being (Anderson and Harrison, 2010; Dewsbury, 2010; Banfield, 2016). In this sense, we can be phenomenologically immersed in the practices of planning just as we can in the practice of daily life. Non-representational thinking therefore encourages us to draw greater equivalence between the affective experiences of residents' practical involvement in their locality and planners' practical involvement in planning. Not only does this establish the similarity between planner concerns over the validity and appropriateness of communities meddling in their professional business and community concerns over planners meddling in their locality, but also suggests similarity between community and planner experiences of the neighbourhood planning process as being either affirming or unsettling. Importantly, if the formal spatial knowledge of planners and the informal place knowledge of communities are not so radically different after all – phenomenologically-speaking – then perhaps there is more scope than we tend to assume to bring those two knowledges together. At this stage, such suggestions are conceptually speculative, but through the chapters that follow, this is one possibility that is explored.

Conclusion

Neighbourhood planning has been described as a journey into the subjectivity of place (Bradley, 2017b), but I would articulate it slightly differently because local people are enrolled in the subjectivity of place on a daily basis. Rather, and reflecting my own disciplinary positioning within non-representational geography, I see neighbourhood planning as an attempt at the explication of that subjectivity of place, an attempt to translate it from a pre-reflective sensation to a communicable idea (Gendlin, 1993, 1995) in a format that might help to preserve that sense of place. The difficulty in doing so is evidenced by the reliance on the term 'sense of place', which has been described as leaping out of nearly every neighbourhood plan (Bradley, 2017b). Like 'genius loci' (Norberg-Schultz, 1980), sense of place is a shorthand term for those special qualities that make a place unique: strongly felt but difficult to pinpoint or define. There is something seemingly intangible, ineffable, about sense of place, such that we recognise its force but struggle to put it into words, let alone locate it on a map. There is an inherent contradiction, then, between the proclaimed opportunity through neighbourhood planning for local communities to protect what makes their area special and the impossibility of doing so through a planning document, as sense of place itself is multiple, mutable and unfinished, and is difficult to put into words and maps. Moreover, the type of words demanded by the planning system is such that any attempt to do so would fail the test of operational applicability because sentiment and sensation are not the business of planning. This is a fundamental problem for neighbourhood planning, which relies on that sense of place to motivate civic engagement, but which can only deal in legally enforceable language and abstract spatial mapping that precludes and risks irreversibly damaging that which it seeks to capture.

Even putting this existential challenge to one side, though, neighbourhood planning does not happen in a vacuum, nor is it simply a matter of finished documents moving through a prescribed and regimented system: neighbourhood planning is performed by people making day-to-day decisions in the context of a whole network of connections, dependencies and responsibilities. While the neighbourhood planning process is prescribed by law and is orchestrated within the structures and protocols of a local authority, this is also not black and white, as some aspects of the process are not specified in detail and the different actors involved are multiply and complexly interconnected. The LPA, for example, has duties to ensure that the local plan is adopted and that it meets its legal obligations, but also has a duty to support neighbourhood planning activity in its area. Developers and site promoters have their financial targets and profit objectives to meet, but in part the LPA's responsibilities in terms of environmental protection, infrastructure provision and affordable housing can only be delivered and secured through effective engagement with the developers/site promoters, whose own financial targets are in part dependent upon the planning process and outcomes. While some community objectives might fit with the environmental agenda of the local authority or the proposed site layout of the developer, others might not, generating a highly contingent and contested arena of practice. There is a performativity to neighbourhood planning, just as there is a performativity to place, and these are recursively related, partially constituting one another. This performativity, then, needs to be taken into account, not least because it perhaps opens the door to greater equivalence being drawn between the spatial experiences of residents and planners, which in turn might bring new opportunities to reduce the incommensurability between these different forms of place knowledge.

These argumentative angles form the bedrock of the chapters that follow, looking first at divergent understandings of place in relation to the environment, the Green Belt and sustainable development in Chapter 3 to expose the fundamental incommensurability of different constructions of place. Subsequently, I consider how neighbourhood planning might be operationalised strategically in Chapter 4 with a shift in focus from the neighbourhood plan as a document to neighbourhood planning as a process, before unearthing the inherently performative aspects of the doing of neighbourhood planning in Chapter 5. Reflecting on the implications of these investigations in Chapter 6, I consider in critical detail the implications of neighbourhood planning for place identity and place attachment, I interrogate issues of legitimacy within neighbourhood planning in its current form, and I evaluate the prospects for neighbourhood planning in both its current and potentially revised form.

References

Agnew, J (2005) Space: place. In PJ Cloke and R Johnston (eds) *Spaces of geographical thought: deconstructing human geography's binaries (1)*. Sage: London, p73–85

Anderson, P and Harrison, P (2010) The promise of non-representational theories. In B Anderson and P Harrison (eds) *Taking place: non-representational geographies and geography*. Ashgate: Farnham, p1–34

Banfield, J (2016) *Geography meets Gendlin: an exploration of disciplinary potential through artistic practice*. Palgrave Macmillan: New York

Beauregard, R (2013) The neglected places of practice. *Planning Theory and Practice* 14 (1): 8–19

Bradley, Q (2017a) Neighbourhoods, communities and the local scale. In S Brownill and Q Bradley (eds) *Localism and neighbourhood planning: power to the people?* Policy Press: Bristol, p39–55

Bradley, Q (2017b) A passion for place: the emotional identifications and empowerment of neighbourhood planning. In S Brownill and Q Bradley (eds) *Localism and neighbourhood planning: power to the people?* Policy Press: Bristol, p163–179

Bradley, Q (2017c) Neighbourhood planning and the impact of place identity on housing development in England. *Planning Theory and Practice* 18 (2): 233–248

Bradley, Q (2018) Neighbourhood planning and the production of spatial knowledge. *The Town Planning Review* 89 (1): 23–42

Bradley, Q and Brownill, S (2017) Reflections of neighbourhood planning: towards a progressive localism. In S Brownill and Q Bradley (eds) *Localism and neighbourhood planning: power to the people?* Policy Press: Bristol, p251–267

Bradley, Q, Burnett, A and Sparling, W (2017) Neighbourhood planning and the spatial practices of localism. In S Brownill and Q Bradley (eds) *Localism and neighbourhood planning: power to the people?* Policy Press: Bristol, p57–74

Brownill, S (2017a) Assembling neighbourhoods: topologies of power and the reshaping of planning. In S Brownill and Q Bradley (eds) *Localism and neighbourhood planning: power to the people?* Policy Press: Bristol, p145–161

Brownill, S (2017b) Neighbourhood planning and the purposes and practices of localism. In S Brownill and Q Bradley (eds) *Localism and neighbourhood planning: power to the people?* Policy Press: Bristol, p19–38

Brownill, S and Bradley, Q (2017) Introduction. In S Brownill and Q Bradley (eds) *Localism and neighbourhood planning: power to the people?* Policy Press: Bristol, p1–15

Cleary, A, Fielding, KS, Bell, SL, Murray, Z and Roiko, S (2017) Exploring potential mechanisms involved in the relationship between eudaimonic wellbeing and nature connection. *Landscape and Urban Planning* 158: 119–128

Colomb, C (2017) Participation and conflict in the formation of neighbourhood areas and forums in 'super-diverse' cities. In S Brownill and Q Bradley (eds) *Localism and neighbourhood planning: power to the people?* Policy Press: Bristol, p127–144

Cresswell, T (2014) *Place: an introduction*. Wiley: Chichester

DCLG (2012) *Neighbourhood planning (general) regulations* www.legislation.gov.uk/uksi/2012/637/contents/made

Dewsbury, J (2010) Language and the event: the unthought of appearing worlds. In B Anderson and P Harrison (eds) *Taking place: non-representational geographies and geography*. Ashgate: Farnham, p147–160

Dorn, ML, Keirns, CC and Del Casino Jnr, VJ (2010) Doubting dualisms. In T Brown, S McLafferty and G Moon (eds) *A companion to health and medical geography*. Wiley-Blackwell: Chichester, p55–78

Fenster, T and Misgav, C (2014) Memory and place in participatory planning. *Planning Theory and Practice* 15 (3): 349–369

Gallent, N and Robinson, S (2013) *Neighbourhood planning: communities, networks and governance*. Policy Press: Bristol

Gendlin, ET (1993) Words can say how they work. In RP Crease (ed) *Proceedings, Heidegger conference*. State University of New York: Stony Brook, p29–35

Gendlin, ET (1995) Crossing and dipping: some terms for approaching the interface between natural understanding and logical formulation. *Minds and Machines* 5: 547–560

Jack, G (2012) The role of place attachments in wellbeing. In S Atkinson, S Fuller and J Painter (eds) *Wellbeing and place*. Routledge: Abingdon, p89–104

Lennon, M and Moore, D (2018) Planning, 'politics' and the production of space: the formulation and application of a framework for examining the micropolitics of community place-making. *Journal of Environmental Policy and Planning* DOI:10.1080/1523908X.2018.1508336

Massey, D (1993) Questions of locality. *Geography* Apr., 142–149

Massey, D (2005) *For space*. Sage: London

Norberg-Schultz, C (1980) *Genius Loci: towards a phenomenology of architecture*. Academy Editions: London

Parker, G (2012) *Neighbourhood planning: precursors, lessons and prospects*. 40th Joint Planning Law Conference, Oxford www.quadrilect.com/Gavin%20Parker.pdf Accessed 15 Dec 2018

Parker, G, Lynn, T and Wargent, M (2015) Sticking to the script? The co-production of neighbourhood planning in England. *The Town Planning Review* 86 (5): 519–536

Parker, G, Lynn, T and Wargent, M (2017) Contestation and conservatism in neighbourhood planning: reconciling agonism and collaboration? *Planning Theory and Practice* 18 (3): 446–465

Parker, G and Salter, K (2017) Taking stock of neighbourhood planning 2011–2016. *Planning Practice and Research* 32 (4): 478–490

Relph, E (1976) *Place and placelessness*. Pion Ltd: London

Stephenson, J (2010) People and place. *Planning Theory and Practice* 11 (1): 9–21

Tuan, Y-F (1974) *Topophilia: a study of environmental perceptions, attitudes and values*. Colombia University Press: New York

Tuan, Y-F (1975) Place – experiential perspective. *Geographical Review* 65: 151–165

van der Pennen, T and Schreuders, H (2014) The fourth way of active citizenship: case studies from the Netherlands. In N Gallent and D Ciaffi (eds) *Community action and planning: contexts, drivers and outcomes*. Policy Press: Bristol, p106–122

Vigar, G, Gunn, S and Brookes, E (2017) Governing our neighbours: participation and conflict in neighbourhood planning. *The Town Planning Review* 88 (4): 423–442

Wargent, M and Parker, G (2018) Re-imagining neighbourhood governance: the future of neighbourhood planning in England. *The Town Planning Review* 89 (4): 379–402

Wills, J (2016) *Locating localism: statecraft, citizenship and democracy*. Policy Press: Bristol

3 Thematic constructions of place

Introduction

The NPPF was introduced in 2012 to simplify the planning system, and to place sustainable development at its heart. It provides a framework within which locally prepared plans can be produced, and local and neighbourhood plans must take it into account. The NPPF was updated in 2018, but sustainable development remains at its heart: 'The purpose of the planning system is to contribute to the achievement of sustainable development' (MHCLG, 2018: paragraph 7), where sustainable development is defined as meeting the needs of the present without compromising the ability of future generations to meet their own needs. Three inter-related thematic objectives underpin this commitment to sustainable development, which the NPPF states need to be pursued in mutually supportive ways to generate net gains: economic, social and environmental. This presumes that economic growth need not be oppositional to either social or environmental protection, and the NPPF establishes a 'presumption in favour of sustainable development', which means that plans should 'positively seek opportunities to meet the development needs of their area' (MHCLG, 2018: paragraph 11).

The NPPF conveys the government's view as to what constitutes sustainable development in the context of planning (Parker et al, 2019), enshrining it within the planning system, so in theory every plan that is adopted in England should contribute to sustainable development as all lower level plans must conform with the NPPF. On the surface, then, the NPPF seems to provide a clear route to eradicate dissensus among actors in the planning system. However, as the NPPF does not render equivalent the three facets of sustainable development (economic, environmental and social) but instead prioritises economic development (Cowell, 2015; Bradley, 2018), the way that sustainable development is framed in the NPPF serves to sustain rather than resolve conflict. As a result, rather than planning being the arena in which deliberation can take place among equals, it is an arena in which deliberation either takes place among interests of unequal power or it does not take place at all.

This elision of political dissensus in planning, though, does not eradicate the political dissensus, as is made clear by a close reading of the construction of core themes that fed into our experience of neighbourhood planning: the environment, the Green Belt and sustainability. It simply hides tensions from sight and prevents them from being aired, let alone resolved. Attending to the different thematic understandings of place that are employed by different actors in the planning system, it becomes clear that these varied actors do not share a common symbolic space but use the same terminology in highly divergent and often flatly contradictory ways, which mitigates against the understandings that underpin neighbourhood planning agendas from gaining traction in the planning system. However, by being aware of these differences in understanding, neighbourhood planning teams can raise these different perspectives within the planning process in a strategic fashion, thereby enhancing the possibility that the implicit conceptual bases and biases upon which planning practice rests might be reconsidered by those with more formal power in the process, raising the possibility of a reconfiguration of potential outcomes in favour of neighbourhood planning agendas. That said, these dividends are not guaranteed, and their success or otherwise depends in large part on the performativity of the neighbourhood planning team in formal settings such as the examination of the local plan and neighbourhood plan, as well as upon the receptivity of those with more formal power in the planning process to such reconsideration, which is explored in more detail in the next chapter.

The National Planning Policy Framework (NPPF)

The NPPF is organised around 13 aims, covering – to varying degrees – the economic, social and environmental objectives of sustainable development, and considering the changes to the NPPF between 2012 and 2018 is informative in this regard (see Table 3.1). Although the general balance between coverage of the three sustainable development objectives is similar between the two time periods, the aim of securing a prosperous rural economy has been dropped in favour of a new aim to make effective use of land, while ensuring the vitality of town centres remains an explicit aim, suggesting either promotion of development in towns or deprioritisation of vitality in rural areas. In 2018, the supply of housing is now the primary aim, rather than a strong, competitive economy, although as housebuilding is a key driver of development, there is much overlap between these two aims and the change perhaps makes little practical difference other than to increase emphasis on house building. Community health and safety is now a higher priority than it was in 2012, but other socially inflected aims, such as transport, communications and design, remain hovering around the middle of the table. However, the explicitly environmental aims still come way down the list. Hence, the primacy of the economy identified earlier can be discerned in the priorities of the NPPF: the economy comes first, then society and finally the environment.

Two specific examples illustrate this prioritisation further. In relation to transport, the NPPF states that applications for development should facilitate access to high quality public transport, '*with layouts that maximise the catchment area for bus or other public transport services'*, and appropriate facilities that encourage public transport use' (MHCLG, 2018, paragraph 110.a, emphasis added). While this requirement for maximising the catchment area for the public transport provider suggests that it also maximises the number of people who can access this

Table 3.1 NPPF Sustainable Development Aims 2012 and 2018

Aim	*NPPF (DCLG, 2012)*	*NPPF (MHCLG, 2018)*
1	Building a strong, competitive economy	Delivering a sufficient supply of homes
2	Ensuring the vitality of town centres	Building a strong competitive economy
3	Supporting a prosperous rural economy	Ensuring the vitality of town centres
4	Promoting sustainable transport	Promoting healthy and safe communities
5	Supporting high quality communications infrastructure	Promoting sustainable transport
6	Delivering a wide choice of high-quality homes	Supporting high-quality communications
7	Requiring good design	Making effective use of land
8	Promoting healthy communities	Achieving well-designed places
9	Protecting Green Belt land	Protecting Green Belt land
10	Meeting the challenge of climate change, flooding and coastal change	Meeting the challenge of climate change, flooding and coastal change
11	Conserving and enhancing the natural environment	Conserving and enhancing the natural environment
12	Conserving and enhancing the historic environment	Conserving and enhancing the historic environment
13	Facilitating the sustainable use of minerals	Facilitating the sustainable use of minerals

service, no consideration is given as to who and where might be differentially affected by this requirement. Seemingly, then, it is only the aggregate customer base that is important, not social equity, and it is only the sustainability of the development being proposed that counts, not the sustainability of existing settlements and communities. Similarly, in relation to designations for environmental protection, the NPPF states that development that is likely to have an adverse effect on a Site of Special Scientific Interest should not be permitted except '*where the benefits of the development in the location proposed clearly outweigh both its likely impact on the features of the site that make it of special scientific interest*, and any broader impacts on the national network of Sites of Special Scientific Interest' (MHCLG, 2018, paragraph 175.b, emphasis added). Even for formally designated sites that are intended to receive particularly high protection from development, it seems that the environment comes first until the economy comes first. While this exception does not define what might count as a 'benefit' of development, this prioritisation of the economy over the environment is a clear demonstration of how terms such as sustainable development, even if framed as a public good, are anything but neutral (Sturzaker and Shaw, 2015). The environmental significance of the site itself is of very limited significance to the determination of whether development is appropriate, and the economic objective of sustainable development always has the potential to take priority.

The same prioritisation can be seen at the intervening level of planning policy – the local plan – as this plan must also conform with the NPPF. In the case of the VWHDC's LPP1 and LPP2 conformity with the NPPF is clear, with commitments made to protect and enhance the built and natural environment, to avoid unacceptably harming the character of the area, respecting the purposes of the Green Belt, and retaining mature trees. However, as with the caveats and exceptions provided in the NPPF, similar 'wriggle-room' is conceivably established through wording such as 'unacceptably harm' and 'respecting'. Who has the authority to determine what constitutes unacceptable harm and against what criteria, and what does it really mean to respect the purposes of the Green Belt both in its own terms and compared to respecting the Green Belt itself? Similarly, as with the NPPF, the economy is prioritised over the environment, with protecting the environment being the last of four themes listed in LPP2 (VWHDC, 2018) around which the planning policies are organised. The NPPF, then, can be considered to deliver a post-political mainstream discourse organised around a notion of a public good (sustainable development) that is difficult to argue against, but which is also difficult to specify, effectively reinforcing a neoliberal pursuit of economic growth above all else, discouraging both the exploration of alternatives (Allmendinger and Haughton, 2010) and debate concerning the appropriateness of prioritising the economic over the social and environmental. Any community that wishes to wrangle over the interpretation and application of the NPPF is obliged to do so through the lower levels of the planning system, but as we will see, this provides no recourse to alternatives either, as different exclusions come into play to prevent a re-politicisation of that which the NPPF serves to depoliticise.

This also raises questions as to the genuine scope for neighbourhood plans even to take precedence on non-strategic matters. Our neighbourhood planning consultation, for example, highlighted the acceptance of the need for some development but also the strength of feeling about the need to maintain and protect the Green Belt, considerable appreciation of locally significant wildlife sites, and support for renewables and eco-builds. Given the lack of clarity as to what it means to respect the purpose of the Green Belt (in LPP2), the potential for harmful impacts on designated wildlife sites to be considered acceptable if other benefits are deemed to outweigh the harm (in the NPPF), and the viability requirements for renewables and eco-builds to be delivered (in the NPPF), the chances of a neighbourhood plan achieving anything tangible seem remote. Even if neighbourhood plans seek to use the NPPF to support their environmental aspirations, and even if their neighbourhood plan is made on this basis, the hierarchy of 'economy first' in the NPPF – and translated into local plans – can always be used in the determination of planning applications to prioritise national or local economic objectives over neighbourhood environmental ambitions. The NPPF, then, is a valuable resource for parties advocating development but arguably an empty vessel for communities seeking to manage that development as appropriate for their locality. Despite providing clear direction in terms of policy priorities, the NPPF can create significant obstacles for the effectiveness of neighbourhood plans in delivering community objectives and serve to embed the conflict between local concerns for the environment and broader concerns for economic growth. In such a scenario, the potential for neighbourhood planning to resolve conflict between stakeholder agendas seems negligible, but the potential for such conflicts to be inflated and further ingrained seems considerable, leaving local communities feeling abandoned, manipulated and aggravated. Hardly conducive to encouraging active citizenship, and a far cry from anticipatory claims that sustainable development would signal a shift from economically to environmentally led planning (English Nature, 1997).

Arguably, then, just as the environmental and the social are simultaneously legitimised by the NPPF (by stipulating that the three pillars of sustainable development be treated in mutually supportive ways) and delegitimised by the NPPF (by relegating them to a position of subservience to economic growth), so the local is at one and the same time legitimised by the Localism Act by granting statutory power to neighbourhood plans and delegitimised by both the Localism Act (by imposing significant constraints on what can be included in such a plan) and by the NPPF (by deprioritising neighbourhood concerns for the environmental quality of their local place relative to broader scale ambitions for economic growth). The potentially deleterious impacts upon neighbourhoods and communities identified in relation to place knowledge, identity and attachment in the previous chapter are thus seemingly re-inscribed here through the stipulations of the NPPF, which in driving the planning system deprioritises the local specificities that neighbourhood planning claims to value and uphold.

It is not only conflict between the economic, social and environmental objectives of sustainable development that causes difficulties for effective neighbourhood planning, though, as even within the field of environmental concerns there is plenty of scope for conflict between stakeholders in neighbourhood planning due to the underlying differences in how key terms are understood, as discussed next in relation to the environment, the Green Belt and sustainability.

Environment

Developing a neighbourhood plan for two rural parishes in an area that is predominantly Green Belt, it is hardly surprising that the environment was a significant concern and priority for local communities. The environment, though, a bit like place, is one of those terms that is both straightforward in an everyday commonsense fashion, and complicated in a formal, scientific and technical fashion. In everyday usage, the environment often refers to green spaces, open countryside, species, habitats and 'environmental issues' such as climate change or pollution. In more formal contexts, the environment can still mean all these things, but it also encompasses concerns for environmental systems as much as green space, such as the atmosphere and the hydrological cycle, species resilience, and the scientific and political definition, measurement, modelling and handling of environmental issues. Often, these different understandings or emphases in discourses about the environment make little or no difference, either because those engaged in the discourse are all speaking from the same understanding or because the differences in understanding are not material to the point being discussed. At other times, though, such differences can be significant. For example, if a formal scientific understanding has determined that to be environmentally significant a patch of green space needs to be above a certain size, then a green space below that size could be expendable for development purposes but on an everyday, common-sense understanding, local communities might not be at all concerned with how large a green space is, as it is a local space that has greenery and species and brings amenity and visual benefits. In such a scenario, not only would development trump environment due to the weighted emphasis embedded in the NPPF, but development would trump environment due to the reliance of planning on formal understandings of scientific definitions, metrics and expertise: if it does not cross the relevant scientific threshold it does not count. In an apparent two-pronged exclusion, then, if the NPPF's perceived prioritisation of the economic over other aspects of sustainability is not sufficient to ensure that development goes ahead, the reliance of planning on technical expertise over lay knowledge (Bradley and Brownill, 2017; Brownill and Bradley, 2017; Parker, 2017; Bradley, 2018; Wargent and Parker, 2018) functions in effect to ensure that the development is realised.

This is a hypothetical scenario, but the written materials produced by different stakeholders in the development of the WSHWNP evidence just how different the underlying understandings of the environment are between the stakeholders. As these understandings are part and parcel of how the same place is constructed differently by those varied stakeholders, it swiftly becomes clear not only how conflict is sustained and magnified rather than resolved by neighbourhood planning but also how significant such conflict is both for the determination of the future for that place and for the implications of that determination for the place identity and attachment of those who live there. Table 3.2 compares the understandings of the environment that are revealed through the emphasis and language of three sets of paperwork relevant to the WSHWNP:

1 The requirements of the Strategic Environmental Assessment Regulations (European Union Directive, 2001/42/EC).
2 The VWHDC Local Plan parts 1 and 2 (VWHDC, 2016, 2018).
3 The results of the year-long public consultation on our neighbourhood plan, contained within our Consultation Statement (WSHWNPSG, 2018).

Several pertinent points arise from this comparison. The Strategic Environmental Assessment (SEA) Regulations are designed to assess, manage and mitigate any environmental impacts likely to arise from development, and a SEA is automatically required if a plan allocates sites for development for housing, employment or retail purposes. In contrast to claims implicit in the NPPF that development is not necessarily oppositional to environmental protection, the automatic requirement for a SEA with any allocation would appear to indicate that development is in fact inherently oppositional to environmental protection, thereby undermining the assumptions underpinning the NPPF.

The SEA Regulations then adopt a scaled approach, whereby the sensitivity or priority of an area or issue influences the severity of the impacts necessary to trigger the need for a SEA (European Union, 2001). In other words, for designated sites, any likely effects arising from development require a SEA but for a general (undesignated) area a SEA is only required if effects arising from development are likely to be significant. The implications of this for the expertise required

Table 3.2 Comparison of understandings of the environment

Strategic Environmental Assessment Regulations	*LPP1/2*	*WSHWNP Public Consultation*
Required if plan allocates land for housing, employment or retail development	Maintain ecological assets where possible; mitigate or compensate for loss where not possible	Greenery, e.g. grass, trees, fields, hedgerows
Also required if likely to affect designated wildlife or conservation areas	Requirement for net biodiversity gain, e.g. installing bird and bat boxes	Designated sites and habitats at risk
Also required if likely adverse impacts on known local environmental issues	Designations and survey duties, e.g. of species and habitats for required SEA assessments	Wildlife, e.g. bats, deer, skylarks, hares
Also required if likely significant environmental effects on an area	Maintain and improve the natural environment, e.g. biodiversity, landscape, green infrastructure, waterways	Environmental issues, e.g. pollution, climate change, flooding
Issues such as fauna, flora, soil, water, air, climatic factors, biodiversity, human health, population, material assets, cultural heritage, landscape and the interrelationship between these factors (these include secondary, cumulative, short, medium and long-term, positive and negative effects)	A list of related policies, e.g.: air quality, waste, external lighting, open space, heritage, contaminated land, advertising public rights of way and conservation	Public rights of way and bridleways

and the workload involved even in determining whether a SEA is required, let alone undertaking the SEA itself, are considerable, with further implications for the quality of the appraisal conducted (Fischer and Yu, 2018). Detailed knowledge is required of the development proposed and its likely environmental effects; the sensitivity of the environment and the presence of explicit designations or issues; the type and level of data required, and the methods used for calculating environmental impacts; and the basis on which evaluations of significance should be made. This is compounded by the extensive list of issues or factors considered to fall under the umbrella of the 'environment' in the SEA Regulations, which is formidable, especially when taken in conjunction with the need to consider their varied temporalities and interactions.

The SEA Regulations, then, adopt an exhaustive, detailed, complicated, impact-based and procedural understanding of the environment which is heavily reliant on formal knowledge and specified technical measures and methods. Looking across the other two columns in Table 3.2, a reduction in complexity and technicality of understandings can be discerned. While the local plan also lists environmental factors that everyday understandings might not recognise (such as advertising, heritage or external lighting) and echoes the need for specific expertise to determine appropriate mitigation and compensation measures, the understanding of the environment evidenced in the local plan is closer to the everyday common-sense understanding. However, when it comes to the public consultation on the neighbourhood plan, there is very little evidence of that formal scientific understanding of the environment. Other than the recognition of local designated sites and habitats, and awareness of global environmental issues, the understanding of the environment here is very much about things that people see around them and – notably – it is about being able to get out into the environment to see those things. Local people are largely aware of the species in their area and their localised presence, even if they do not know whether each one is vulnerable or endangered, and they are aware of the diversity of greenery in their area even if they do not know the exact scientific definition of each habitat. While some residents are very well informed on environmental science and its local significance, many value the environment around them not because they can name it, measure it and evaluate it, but simply because it is there.

Unsurprisingly, then, Table 3.2 indicates a spectrum from formal scientific understandings of the environment in the SEA Regulations to everyday common-sense understandings in the public consultation on the WSHWNP, with the local plan merging these two in an everyday formal understanding, entirely consistent with their need to balance their own requirements under the SEA with their need to reflect local concerns and communicate their proposals to their public. However, what this confirms is that our earlier hypothetical scenario is not hypothetical but real. Beyond any locally designated sites, the Designated Area for the neighbourhood plan is simply 'an area' in SEA terms, so to trigger an SEA any impacts on the Designated Area arising from development proposals would need to be significant, whereas it is probable that any impact on the environment of the Designated Area would be of concern to local residents because it would affect their daily experience of that environment. This is not just a matter of scale, but of the criteria against which significance is assessed: impacts on the functioning of the environment in scientific ecological terms versus impacts on the visible presence of the environment in everyday terms. Similarly, residents might care very much about the potential loss of certain species but care very little about the net biodiversity gain that is the priority for the local plan. This is not to say that local residents would not support the net gain required by a local plan but that they might not prioritise this if it meant losing – for example – the skylarks that are known to breed in the area and which fill the skies with their chirrupy twittering. Net biodiversity gain might come a poor second to protecting specific locally valued species, whether formally 'protected' or not, especially if those species are considered to contribute to the character of the area or place attachment.

These different understandings, then, are fundamental to some of the conflicts that arise in neighbourhood planning: while the net gain requirement of a local plan might be laudable in and of itself when abstracted from the local situation in which it comes into effect, it is that very local situation that forms the basis of a neighbourhood plan. The net gain requirement is a mechanism by which development promoters can legitimise the loss or degradation of certain species and habitats by providing a net gain in biodiversity through other means and/or in another place, which the local community might not care for at all. As such, it is another example of how higher-level plans and formal scientific understandings – and therefore economic development – can reliably and routinely override lower level plans, local everyday understandings and environmental concerns. It assumes that something similar but different and somewhere other than where it originally was has the same worth (value) as the original that it replaces. If the replacement is bigger or more diverse, then there is deemed to be a net gain irrespective of what particularities might have been lost in the process. The environmental 'asset' is thus amputated from its spatial and social setting and is re-established in a 'new and improved' format elsewhere, but it is precisely the emplacement of this environmental asset within its spatial and social setting that instils such value and attachment on the part of residents in the first place. As with place identity and attachment in the previous chapter, then, the same local knowledge and feeling that neighbourhood planning seeks to capture and legitimise is excluded from the very process that seeks to capture and legitimise it.

Green Belt

While the NPPF (2012/18) seems clear in terms of the importance, purpose and permanence of the Green Belt, how this is interpreted in lower levels of the planning system is somewhat murkier, generating another crucible of conflict in neighbourhood planning. The NPPF states that the 'fundamental aim of Green Belt policy is to prevent urban sprawl by keeping land permanently open' (MHCLG, 2018, paragraph 133), and specifies five purposes of the Green Belt which aim to check unrestricted sprawl of large built-up areas, prevent the merging of neighbouring towns, to safeguard the countryside from encroachment, to preserve historic towns and to assist in urban regeneration (MHCLG, 2018; paragraph 134).

As with the environment, issues of definition, measurement and context come into play here as judgements are needed in relation to determining what makes sprawl 'unrestricted', what constitutes a 'large' built-up area, what counts as 'encroachment' and what defines the 'setting and special character' of historic towns. Importantly in the context of the WSHWNP, the use of the word 'town' took on extra significance, as the settlements affected by the change in the allocation of development from LPP1 to LPP2 were villages, not towns. In fact, Whitecross does not even feature on the settlement hierarchy of the local plan as a named settlement at all, and therefore is deemed to be open countryside (VWHDC, 2016). To illustrate this issue, the differences in the underlying understandings in relation to the Green Belt are presented in Table 3.3, which compares the representations of the Green Belt in three sets of paperwork relevant to the WSHWNP:

Table 3.3 Comparison of understandings of the Green Belt

LPP1/2	*Green Belt study*	*WSHWNP public consultation*
Large towns only	Large towns only	Small towns and villages too
Openness and permanence to be protected	Emphasis on openness and permanence	Personal significance
Commitment to respect the purposes of the Green Belt	Can build within the Green Belt if likely harm is outweighed by benefits	A local asset and quality of the environment
LPP1: development to be focused in inset locations and within the built area	Parcels of land assessed for their contribution to visual amenity and settlement separation	Vulnerable to erosion; needs protecting
LPP2: 'exceptional circumstances' to delete land at Dalton Barracks along with adjacent fields and villages	Tree belts as separating parcels of land from the wider landscape	Green views important to character of small, discrete settlements
Providing for Oxford's unmet need and availability of MoD site the 'exceptional circumstances'	Hierarchy of 'parcel' contribution to purposes of Green Belt	Dog walking, accessibility

1 The VWHDC Local Plan parts 1 and 2 (VWHDC, 2016, 2018).
2 A Green Belt study commissioned by VWHDC to explore the impacts of the deletion of land from the Green Belt for development purposes (Hankinson Duckett Associates, 2017).
3 The results of the year-long public consultation on our neighbourhood plan contained within our Consultation Statement (WSHWNPSG, 2018).

Unsurprisingly, there is considerable consistency between the local plan and the Green Belt study, as the Green Belt study was commissioned by the LPA. The local plan's commitment to respect the purposes of the Green Belt rather than the Green Belt itself is significant here, as it conceivably introduces the potential to chip away at areas of the Green Belt deemed to make less of a contribution to those purposes, and the Green Belt study evaluated parcels of land within the Green Belt in terms of their contribution to Green Belt purposes (Hankinson Duckett Associates, 2017). While the NPPF does not explicitly state that the purposes can be seen in graduated terms, it also does not preclude such a graduated conceptualisation of the Green Belt, so there is nothing illicit in this. Interestingly, though, the methodology employed in the Green Belt Study – parcelling up the Green Belt for evaluation – itself potentially runs against one of the defining features of the Green Belt: its openness. The use of tree lines in demarcating one 'parcel' of land from another suggests that the Green Belt is piecemeal rather than continuous as the tree lines are construed as dividing the Green Belt rather than simply being a feature of it, connecting as much as separating the fields on either side. Somewhat ironically, it seems to me that the openness of the Green Belt was assessed only after the Green Belt had been analytically subdivided into parcels – a term that itself suggests enclosure – thereby presenting as a pre-existing feature the outcome that was desired.

Table 3.3 suggests that in general terms and consistent with its framing in the NPPF, the local plan frames the Green Belt as a constraint on development, which can only be overcome if exceptional circumstances can be established to justify deleting land from the Green Belt for development purposes. However, the Green Belt study effectively enables it to be framed as a potential resource if parcels of land within the Green Belt can be shown to contribute less effectively to the purposes of the Green Belt. This perceived process of division and metrication thus seemingly allows the local plan to respect the purposes of the Green Belt but not the Green Belt itself as a singular entity by rendering different parts of the Green Belt more or less important. By contrast, residents do not draw on differential notions of Green Belt purposes but value the Green Belt simply because it is there. They recognise the role that the Green Belt plays in controlling development and protecting the rural and open character of their area and, because they do not tend to consider the different ways in and degrees to which different areas contribute to its purposes, they see the Green Belt in more uniform terms as an essential and valued quality of their area. It is also clear from the public consultation on the WSHWNP that the Green Belt is not simply about land that is currently or might be developable but is about access to, identification with and active engagement in the Green Belt (WSHWNPSG, 2018). Local residents do not simply look at the distribution of the Green Belt on a map and they do not critique its formal contribution to its aims: they live in the Green Belt, and they live in settlements that gain much of their character from the Green Belt and the character of which is in turn protected by the Green Belt. Local support for the Green Belt is often perceived as NIMBYism and an outright rejection of any form or level of development, but in this case this is inaccurate, as it is not development itself that is the focus of community resistance but the predicted impact of that development on the character and perceived quality of their settlement and its environs. Residents appreciate the role that the Green Belt plays in establishing and protecting their lived space, and they perceive that such a designation is the only thing that does protect that character and perceived quality. For residents, then, the Green Belt is singular and absolute, not differentiated and graduated.

The decision on the part of the LPA to accept a proportion of Oxford's unmet housing need and to allocate that development to Green Belt land is especially interesting given that Green Belt designation is valid grounds for declining to cooperate on such matters under the terms of the Localism Act 2011 (DCLG, 2011). Personally, I find it inexplicable that the city cannot build at its edges because of Green Belt constraints that prevent urban sprawl but development for the city can leapfrog over the most immediate Green Belt and impose itself on Green Belt land slightly further away. To me, this leapfrogging of development is simply a different form of sprawl that increases transport pressures by locating accommodation for the city away from the city that it is intended to serve, making it anything but sustainable irrespective of any locally felt injustice. The resistance voiced locally is not (generally-speaking) resistance to development in principle and is certainly not resistance to development within the proposed Strategic Development Site at Dalton Barracks and Abingdon Airfield, which received broad support in the WSHWNP consultation despite the site being in the Green Belt (WSHWNPSG, 2018). Instead, residents are resistant to the deletion of land from the Green Belt as an easy way of delivering what is perceived as unnecessary and excessive development that is intended not to meet local need but to fuel economic growth elsewhere. This is not a case of NIMBYism, then, but of consideration of what constitutes an appropriate scale and location of development given the local social, economic and environmental conditions.

In reality local residents give considerably more thought to strategic issues of how much of what type of development should take place where than the caricature of the NIMBY suggests, yet there seemingly remains a fundamental difference in perspective on the Green Belt around which these strategic considerations crystallise: whereas LPP2 respects the purposes of the Green Belt in a piecemeal fashion, local people respect the Green Belt in totality. By derogating residents as NIMBYs, those in favour of development and/or not in favour of taking local views into account preclude any substantive political discussion of strategic alternatives such as different development models in an antagonistic sense. Simultaneously, the different understandings of concepts such as the environment and the Green Belt prevent any agonistic collaboration in planning because there exists no shared symbolic framework. We might use the same words and phrases, but we use them very differently, and whichever interpretation is favoured in the determination of planning outcomes has significant implications for the future of the place that is multiply and variously constituted by those understandings.

Scalar differences are also clear from Table 3.3. In recognising the role of the Green Belt in preventing only (larger) towns from merging, the way is left clear for the merging of smaller towns and villages. Whereas the Vale of White Horse lacks any explicit cities, it does host several large towns such as Didcot, Abingdon and Faringdon, but also hosts countless small villages as well as hamlets too small to feature within the settlement hierarchy of the local plan, such as Whitecross. Although there are differences in settlement size within the Designated Area, made most apparent through the uneven availability of shops and community facilities, these are less extreme than across the whole district, and in terms of daily life, residents generally have no cause to think formally about settlement hierarchy. In this context, residents apply the purposes of the Green Belt to the settlements within their area irrespective of their formal designation as different types of settlement, whereas the local plan applies the purposes of the Green Belt to towns only.

Table 3.3, then, suggests that for both the local plan and the Green Belt study, merging villages is fine but merging towns is not, whereas for residents, merging settlements is not fine, whether town, village or hamlet. As with the environment, this is not simply a matter of scale, as it raises questions as to whether the NPPF's policy on the Green Belt is about the specific wording of the policy or the spirit of the policy. For those promoting development, it appears to be about the specific wording, so providing that the towns of Abingdon and Wootton did not merge, everything was fine. For those who – while supportive of development at the site in principle – resisted the merging of their settlements with the new development, it was all about the spirit of the policy, so the new development should not be merged with any settlement. These differences make inevitable the continuance of conflict between stakeholders in the absence of appreciation and negotiation of these divergent underlying understandings. The planning system presupposes that all parties conceptualise the Green Belt in a consistent manner, but this is not the case, and to assume that just because residents are being brought into the planning system they will adopt the established professional understandings of planning is mistaken. Instead, they bring their everyday common-sense understandings into planning, and it should fall to the planning system to adapt itself to accommodate these alternative understandings if residents are to be present to the planning system in anything other than body. Without such adaptation, no progress can be made in winning over hearts and minds in support of development through the supposed growth in active citizenship. Rather than the shared symbolic space deemed necessary for collaborative or deliberative planning, we have instead a discursive space of symbolic commonalities that is in practice colonised and dominated by specific, technical, reductive and abstract knowledges (Mouffe, 2005; Parker et al, 2017; Kenis, 2018; Bradley, 2018).

Sustainability

At the outset of this chapter, I explored the different emphasis placed on the three objectives of sustainable development – economic, social and environmental – within the NPPF, but here my concern is with the idea of sustainable development more generally, how it is operationalised in planning documents and how it is understood by residents. As with the environment and the Green Belt, underlying differences in understandings of sustainable development have serious implications for the resolution of conflict and the effectiveness of neighbourhood planning in achieving community objectives. In Table 3.4, I compare the accounts and descriptions of sustainable development between three sets of paperwork relevant to the WSHWNP:

1. The VWHDC Local Plan parts 1 and 2 (VWHDC, 2016, 2018).
2. A Green Belt study commissioned by VWHDC to explore the impacts of the deletion of land from the Green Belt for development purposes (Hankinson Duckett Associates, 2017).
3. The results of the year-long public consultation on our neighbourhood plan contained within our Consultation Statement (WSHWNP, 2018).

One thing that jumps out from this comparison is the distinction between the three objectives of sustainable development identified in the NPPF (economy, society and environment) and the four themes around which the local plan policies

Table 3.4 Comparison of understandings of sustainability

LPP1/2	*Green Belt study*	*WSHWNP public consultation*
NPPF presumption in favour of sustainable development in Core Policy 1	Not an explicit concern of the study	Not clearly defined, but sustainability of development desired
Delivered through four themes: 1 Communities – housing, facilities, accessibility, open space 2 Economy – employment land, innovation, key sites/employers, skills, rural economy 3 Transport – roads, buses, homes/jobs co-located 4 Environment – low carbon, biodiversity, water, landscape	Identified as a driver for Green Belt reviews Notable emphases: 1 Transport and infrastructure 2 Intrinsic qualities of land 3 Visual containment and landscape sensitivity	Focus on sustainability of rural character of settlements and existing communities Associated with desire to keep the area as it is Transport a major sustainability concern: road pressures, bus services, foot and cycle paths
Focus on most sustainable locations: integrated with environment and with adequate facilities, directed by spatial strategy, e.g. market towns	Need further work to assess what is sustainable in this area	Sceptical/suspicious, especially around infrastructure and transport

are organised (communities, economy, transport and environment), immediately indicating that the inter-relatedness of the NPPF objectives makes them amenable to being reconfigured. This is supported by the arrangement of policies in the WSHWNP which – despite needing to conform to both the NPPF and LPP1/2 – are organised as a spatial strategy, a set of infrastructure needs policies and a design guide. Clearly, there is nothing essential or self-explanatory about the definition of the objectives of sustainable development; they are malleable in terms of how they are used to organise policies in lower level plans, even if not in terms of their relative priority.

Another notable feature of the understanding of sustainable development in the local plan is the emphasis on the sustainability of locations, and especially on the relative pre-existing sustainability of different locations (VWHDC, 2016, 2018). While the articulated intention to focus development in areas already considered to be most sustainable might seem sensible on first reading, it risks overlooking areas that might need or benefit from enhanced sustainability through targeted location of appropriate development. In what could be seen as a classic case of the sustainability trap (Sturzaker and Shaw, 2015), settlements deemed too small to be sustainable are potentially deprived of any opportunity to develop even if their residents are in favour of development, while those that are big enough to be deemed sustainable are seemingly assumed to become even more so if they become even bigger. While some neighbourhood plans have successfully overturned settlement hierarchies in the local plan to permit greater levels of development in smaller settlements (Bradley, 2017), any tendency to equate larger sized settlements with greater sustainability directly contradicts the mantra of the localist agenda that assumes that the small and local is the best and natural scale at which to enact planning and governance (Madanipour and Davoudi, 2015; Cowie and Davoudi, 2015; Parker and Salter, 2017), raising doubts as to the deliverability of sustainable development at the neighbourhood level if neighbourhood planning is done at the small scale but sustainability is a feature only of the large scale.

The Green Belt study is markedly different in this regard, as it seems to recognise the potential for sustainable development to act as a force, rather than being a quality of a (large) settlement (Hankinson Duckett Associates, 2017). While most residents would not wish to see sustainable development used as a motivation for removing land from the Green Belt, the idea of sustainable development as a force or motivation might lend a different or at least a more diverse approach to sustainable development within planning locally. This would then be more in keeping with the sensitivities evident in the public consultation on the neighbourhood plan which indicated an informal and implicit understanding of sustainability rather than sustainable development, emphasising the sustainability of all settlements (not just new developments) and – notably – the sustainability of the *character* of settlements, which links sustainability directly to the protection of the Green Belt. While residents might struggle to articulate a detailed and formal understanding of sustainable development as the local plan does, they are alert to the social injustice that they perceive in development proposals that deliver 'sustainable' new developments at the expense of existing settlements and communities. This is especially pertinent in relation to transport, as early proposals in LPP2 (now deleted) included constructing bus and cycle routes across the Green Belt to a new Park and Ride facility with seemingly no undertaking to make these services optimally available for existing settlements near the route: it would cut across a linear village, thereby adding to congestion and pollution but seemingly not providing a service to that settlement, and by-pass a village that currently lacks public transport (VWHDC, 2018). Similarly, in the early stages of master planning for the Strategic Development Site, it was proposed that a premium (four times per hour) direct bus service would be provided by diverting the service away from its current route (DIO, 2017). Although explicit reference to this diversion has since been deleted, this would halve the service on part of the existing route, evidencing again the prioritisation of sustainability for new developments at the expense of existing settlements, with such service revisions deemed by other statutory bodies in their responses to the neighbourhood plan consultation as being inevitable with such a large-scale development (WSHWNPSG, 2018).

As with the environment and the Green Belt, then, sustainable development is not understood in the same way by the different parties, which complicates negotiations and ensures that no genuine consensus can be achieved. Debate simply falls back on oppositional terms between concerns for the economy on the one hand and for communities and the environment on the other. Furthermore, the dismissal of neighbourhood concerns about loss of Green Belt, environmental mitigation and sustainability as reactionary NIMBYism regarding parochial details does a considerable disservice to local residents, whose consideration of the issues at stake is far deeper, more thorough-going and more integrated than is often acknowledged. It also denies the institutional parochialism that enforces a professionally blinkered understanding of the Green Belt, the environment and sustainable development as a form of epistemic violence that further excludes local place knowledge and priorities (Madanipour and Davoudi, 2015; Parker et al, 2017; Bradley, 2018). Environmental issues, the Green Belt and sustainability are interlinked concerns that are implicated in settlement character, place identity and place attachment. They are therefore deeply embedded in the very place attachment upon which neighbourhood planning is predicated. Consequently, the repeated denial and dismissal of local knowledge in relation to a whole host of concepts and issues that contribute to the construction of place undermines the stated intentions of neighbourhood planning to legitimise, involve, protect and nurture local knowledge and place attachment. As the validity of local knowledge is progressively dismantled, the strength and positive valence of place attachment also comes under fire, potentially rendering neighbourhood planning not only self-limiting but self-defeating as it serves to devour the very place attachment upon which it rests.

Conclusion

This chapter shows that although the purposes of the NPPF are to simplify the planning process, provide clear direction for lower level plans, and put sustainable development at the heart of the planning system, these objectives are anything but seamlessly delivered. Far from evidencing that economic development does not have to be at the expense of social and environmental protection or enhancement, my reading of the NPPF reinforces the view that the economic will always trump the social and the environmental, such that although the three objectives are not necessarily conflicting, where they do conflict, it is the economy that will be the victor. While all three objectives of sustainable development are addressed in the NPPF, they are not addressed equally nor are they valued equally. One could easily argue, on this basis, that the presumption in favour of sustainable development has simply morphed into a presumption that all development is sustainable and that as neighbourhood plans cannot refuse development allocations, any grounds that a neighbourhood planning team thinks it can find in the NPPF to support its aims will rapidly turn to quicksand.

This perpetual prioritisation of the economic side of sustainability above the social and the environmental perceived in the NPPF is unassailable in practice as the NPPF is the driving force behind planning in England, even though in our case it seemingly contravenes the aim of the national sustainable development strategy to meet 'the diverse needs of all people in existing and future communities' (DEFRA, 2005: 16) by deprioritising existing communities. Equally, close inspection of different understandings of sustainable development highlighted the selective impacts of directing development to already sustainable areas at the expense of using development to enhance the sustainability of less sustainable areas, and of prioritising the sustainability of new developments at the expense of the sustainability of existing settlements. Pursuing net gains is neither sufficient nor appropriate to accommodate the needs, concerns, aspirations and objectives of local people, because their perspective is more local, more informal and more specific than the higher levels of planning, but no less legitimate for that.

The imbalance in priority afforded to concerns that are likely to be paramount at different levels in the planning system and on the part of different stakeholders both leave unaddressed and actively sustain the conflicts that arise as a result of divergent underlying understandings of terms or themes such as the environment, the Green Belt and sustainable development. In each case, the perspective of residents was markedly different to those of the other stakeholders considered and even when residents are supportive in principle to the approaches adopted (such as net biodiversity gain), their enhanced sensitivity to the specificities of the circumstances in which those approaches come to bear makes them resistant in practice. This is not simply a matter of scale, but a matter of the criteria used in evaluating the meaning, significance or value of the environment or the Green Belt, based not on formal scientific knowledge, metrics and expertise but on local lived experience and attachment. Green spaces are valued not because of their contribution to the functional resilience of the ecosystem but because they contribute to the character and quality of the area, to what it means to live in that place. Equally, skylarks are cherished not (just) because they are protected but simply because they are there and because they contribute to the experiential character of the environment in which local people live, and to place identity and place attachment. These contributions, though, are excluded from planning and in the absence of any meaningful attempt to incorporate these forms of knowledge into planning, local participation in planning will only ever be partial, marginal and minimal, falling well short of the aims of localism and bringing detrimental implications for its long-term potential. While the emergence of Latourian notions of multi-naturalism, whereby different natures are brought into being by different constellations of actors, might help us to conceptualise the divergent and incompatible constructions of the environment and of place identified in this chapter, recognising these multiple and contested knowledges, natures and places does nothing to resolve the contestation itself (Lorimer, 2012; Freitag, 2018). It does not remove the political nature of planning but rather emphasises the need to re-politicise planning.

For planning to be properly political, the antagonism that underpins society needs to be acknowledged rather than denied or masked, and for planning to be properly collaborative or deliberative it must take place within an agonistic discursive space of shared symbolism, in which conflicting parties recognise the legitimacy of their opponent's positions and perspectives (Mouffe, 2005; Kenis, 2018). This is simply not the case in neighbourhood planning because communities are denied, excluded and delegitimised as their voices are considered dissident from the mainstream pursuit of economic growth at all costs (Allmendinger and Haughton, 2010; Etherington and Jones, 2017). Despite the optimistic claims of some that neighbourhood planning might constitute a step towards more collaborative planning in which a way of being comes into dialogue with a way of knowing (Parker, 2014; Bradley, 2018), to the extent that the resolution of conflict and tension only seemingly becomes feasible by the eradication of neighbourhood agonists from the deliberations, neighbourhood planning can in no way be considered collaborative or deliberative in anything other than a fictional sense. We might be invited to speak, but our voices can be immediately silenced. While the Localism Act aims to multiply and democratise the planning system, neighbourhood planning does not – and more importantly, I suggest, cannot – deliver this multiplication and democratisation in its current form because while legitimising local knowledge with one hand, it simultaneously delegitimises that same local knowledge with the other. In the first instance, the operationalisation of sustainable development identified in the NPPF ensures that local environmental and social concerns are always the lowest priority. Secondly, the alternative futures that constitute the political are hidden from view by the enforced focus at the neighbourhood level on technicalities and specificities rather than anything strategic, so antagonistic debate is precluded. Third, even with the best will in the world on the part of those involved, the abject lack of a shared symbolic framework upon which to base agonistic deliberative or collaborative planning prevents any meaningful incorporation of local knowledge into planning.

The challenge then becomes how to reinsert local knowledge and concerns into the very process that excludes you even while claiming to legitimise you, in such a way that local concerns are re-legitimised in an unconventional manner. This, naturally, renders any collaborative approach to planning even less likely to be effective than was already the case, as the conventional parameters for the involvement of neighbourhood planning teams in the broader planning system are progressively challenged and transgressed. It is at this point that the strategic potential of neighbourhood planning opens up, but it only does so through this unconventional reconfiguration of what it means to do neighbourhood planning, and it is this issue to which my attention turns in the next chapter.

References

Allmendinger, P and Haughton, G (2010) Spatial planning, devolution, and new planning spaces. *Environment and Planning C: Government and Policy* 28: 803–818

Bradley, Q (2017) Neighbourhood planning and the impact of place identity on housing development in England. *Planning Theory and Practice* 18 (2): 233–248

Bradley, Q (2018) Neighbourhood planning and the production of spatial knowledge. *The Town Planning Review* 89 (1): 23–42

Bradley, Q and Brownill, S (2017) Reflections of neighbourhood planning: towards a progressive localism. In S Brownill and Q Bradley (eds) *Localism and neighbourhood planning: power to the people?* Policy Press: Bristol, p251–267

Brownill, S (2017) Assembling neighbourhoods: topologies of power and the reshaping of planning. In S Brownill and Q Bradley (eds) *Localism and neighbourhood planning: power to the people?* Policy Press: Bristol, p145–161

Brownill, S and Bradley, Q (2017) Introduction. In S Brownill and Q Bradley (eds) *Localism and neighbourhood planning: power to the people?* Policy Press: Bristol, p1–15

Cowell, R (2015) Localism and the environment: effective rescaling for sustainability transition. In S Davoudi and A Madanipour (eds) *Reconsidering localism*. Routledge: New York and London, chapter 11. (no page numbers) Accessed 6 Dec 2018

Cowie, P and Davoudi, S (2015) Is small really beautiful? The legitimacy of neighbourhood planning. In S Davoudi and A Madanipour (eds) *Reconsidering localism*. Routledge: New York and London, chapter 9. (no page numbers) Accessed 6 Dec 2018

DCLG (2012) *National planning policy framework* www.gov.uk/government/publications/national-planning-policy-framework-2

DCLG (2011) *The localism act* www.legislation.gov.uk/ukpga/2011/20/contents/enacted

DEFRA (2005) *Securing the future: delivering UK sustainable development strategy* www.gov.uk/government/publications/securing-the-future-delivering-uk-sustainable-development-strategy

DIO (2017) *Dalton Barracks delivery document* www.whitehorsedc.gov.uk/java/support/dynamic_serve.jsp?ID=902932729&CODE=146712F47A0E0153F7A2FA74D0EA34AF Accessed 23 Jan 2019

English Nature (1997) *A framework for the future: green networks with multiple uses in and around towns and cities*. English Nature research report No 256. file:///F:/green%20networks%20report.pdf

Etherington, D and Jones, M (2017) Re-stating the post-political: depoliticization, social inequalities, and city-region growth. *Environment and Planning A: Economy and Space* 50 (1): 51–72

European Union (2001) *Strategic environmental assessment regulations* (European Directive 2001/42/EC)

Fischer, TB and Yu, X (2018) Sustainability appraisal in neighbourhood planning in England. *Journal of Environmental Planning and Management*. DOI:10.1080/09640568.2018.1454304

Freitag, A (2018) Visions of wilderness in the North Bay communities of California. *Area* 50 (1): 91–100

Hankinson Duckett Associates (2017) *Green Belt study – land surrounding Dalton Barracks* Available via LPP2 public examination library www.whitehorsedc.gov.uk/services-and-advice/planning-and-building/planning-policy/local-plan-2031-part-2

Kenis, A (2018) Post-politics contested: why multiple voices on climate change do not equal politicisation. *Environment and Planning C: Politics and Space* DOI:10.1177/0263774X18807209

Lorimer, J (2012) Multinatural geographies for the anthropocene. *Progress in Human Geography* 36: 593–612

Madanipour, A and Davoudi, S (2015) Localism: institutions, territories, representation. In S Davoudi and A Madanipour (eds) *Reconsidering localism*. Routledge: New York and London, chapter 2. (no page numbers) Accessed 6 Dec 2018

MHCLG (2018) *National planning policy framework* www.gov.uk/government/publications/national-planning-policy-framework-2

Mouffe, C (2005) *On the political*. Routledge: Abingdon

Parker, G (2014) Engaging neighbourhoods: experiences of transactive planning with communities in England. In N Gallent and D Ciaffi (eds) *Community action and planning: contexts, drivers and outcomes*. Policy Press: Bristol, p177–200

Parker, G (2017) The uneven geographies of neighbourhood planning in England. In S Brownill and Q Bradley (eds) *Localism and neighbourhood planning: power to the people?* Policy Press: Bristol, p75–91

Parker, G, Lynn, T and Wargent, M (2017) Contestation and conservatism in neighbourhood planning: reconciling agonism and collaboration? *Planning Theory and Practice* 18 (3): 446–465

Parker, G and Salter, K (2017) Taking stock of neighbourhood planning 2011–2016. *Planning Practice and Research* 32 (4): 478–490

Parker, G, Salter, K and Wargent, M (2019) *Neighbourhood planning in practice*. Lund Humphries: London

Parker, G and Street, E (2015) Planning at the neighbourhood scale: localism, dialogical politics, and the modulation of community action. *Environment and Planning C: Government and Policy* 33: 794–810

Sturzaker, J and Shaw, D (2015) Localism in practice: lessons from a pioneer neighbourhood plan in England. *The Town Planning Review* 86 (5): 588–609

Vigar, G, Gunna, S and Brookes, E (2017) Governing our neighbours: participation and conflict in neighbourhood planning. *The Town Planning Review* 88 (4): 423–442

VWHDC (2016) *Vale of white horse district council local plan 2031, part 1: strategic sites and policies* www.whitehorsedc.gov.uk/services-and-advice/planning-and-building/planning-policy/new-local-plan-2031-part-1-strategic-sites

VWHDC (2018) *Vale of white horse district council local plan 2031, part 2: detailed policies and additional sites* www.whitehorsedc.gov.uk/services-and-advice/planning-and-building/planning-policy/local-plan-2031-part-2

Wargent, M and Parker, G (2018) Re-imagining neighbourhood governance: the future of neighbourhood planning in England. *The Town Planning Review* 89 (4): 379–402

WSHWNPSG (2018) *Wootton and St Helen Without Neighbourhood Plan 2018–2031. consultation statement* www.whitehorsedc.gov.uk/sites/default/files/WSHWNP%20Consultation%20Statement_0.pdf

4 Strategic constructions of place

Introduction

It is not unusual in planning literature to read concerns about the ability of residents to articulate their aspirations and objectives in terms of strategy or principle, the need to translate these aspirations and concerns into the technical language of planning, and the loss of neighbourhood meaning and sentiment along the way (see, for example, Gallent and Robinson, 2013; Parker et al, 2015; Wargent and Parker, 2018). It could be argued that these are two separate sets of concerns: one related to the capacity to think strategically and the other related to the need to write technically. However, they are very much interconnected as technically articulated planning policies tend to convey broad-brush in-principle intentions and permissions, along the lines of 'development will be permitted if . . . ' statements, which are perhaps followed by a string of exceptions, conditions and caveats. Supplementing these principled statements are more specific policies directed towards individual locations, such as Strategic Development Sites that are allocated for long-term use. Immediately, then, we encounter two uses of the word 'strategic': the first as a way of communicating generic, non-specific permissions across large areas; the second as a policy statement denoting long-term intentions targeted towards specific sites. Interestingly, the planning policies that are applied to a Strategic Development Site are not themselves strategic in the first sense of the word as they are targeted rather than non-specific, risking the establishment of policies that diverge from or directly contradict the in-principle policies.

This complex relationship between the two meanings of strategy helps to unpick some of the intractable difficulties inherent to neighbourhood planning that undermine its own potential and further entrench conflict between the different actors involved. For one thing, neighbourhood plans are explicitly prohibited from addressing strategic matters, but what form of 'strategic' applies here? Neighbourhood plans are permitted to allocate additional development sites, and if this is intended to direct development in the longer term, how different is this from the strategic allocations of a local plan? Surely this is just a matter of scale, rather than a distinction between being strategic or not. This issue of the strategic being a matter of scale is addressed in the first section of this chapter, which unsettles any clear distinction between a local plan being strategic and a neighbourhood plan not being strategic. Secondly, compared to the area covered by a local plan, a neighbourhood plan might cover a spatial extent no greater than a Strategic Development Site, so if a policy in such a plan relates to the whole Designated Area for that neighbourhood plan, would this constitute a strategic policy in the sense of an in-principle statement across the whole area, a strategic policy for a specific site in the sense of long-term intentions, or both? This issue of strategy being explicitly spatialised is the focus of the second section of this chapter, which proposes that it is more a matter of perspective than the nature or level of a planning document that determines how strategy is implemented. Finally, if being strategic can apply to the longer term as well as the larger scale, then what is to stop neighbourhood planning teams from developing a temporal strategy in relation to the development of their plan to supplement the development of a spatial strategy in relation to the place of their own neighbourhood? The opportunities and difficulties that we encountered in relation to this way of being strategic are explored in the third and final section of this chapter.

Through this discussion, I establish an argument that concern focused on the ability or otherwise of local residents to articulate their aspirations in terms of strategy or principle somewhat misrepresents the challenge for neighbourhood planners in four keys ways: 1) it over-emphasises the first meaning of strategy at the expense of the second meaning, 2) it does not consider sufficiently the implications of the restrictions placed upon neighbourhood plans in terms of their scope to be strategic, 3) it overlooks the interaction between the spatial in-principle and temporal intentional meanings of strategy, and 4) it leaves unaddressed both opportunities for neighbourhood planning to act in a more strategic fashion than is often assumed and the difficulties and implications of doing so.

Strategy as scalar: questioning the strategic

The neighbourhood planning regulations and guidance make it clear that strategy is done at the local level and neighbourhood plans do not do strategy, they only conform with the local level strategy (Gallent and Robinson, 2013; Gallent, 2014; Parker, 2014; Wills, 2016; Bradley and Brownill, 2017; Brownill and Bradley, 2017; Parker et al, 2019). A local plan is a strategic document, laying out the proposals for the management of development in a local authority area in terms of location, scale, infrastructure requirements and environmental protection. A neighbourhood plan must be in broad conformity with, and must not contravene the strategic objectives of, the relevant local plan. The strategic aspects of a neighbourhood plan are therefore specified in advance by the local plan to which it relates. However, in practice this distinction is easily unsettled as LPAs do not always stick to the script when it comes to confining strategic intentions to strategic documents, there is considerable uncertainty as to what constitutes strategy, and neighbourhood planning teams are equally capable of acting strategically even if they cannot directly determine the strategy that is set for their area.

Specifically, areas covered by a neighbourhood plan cannot refuse strategic housing allocations specified in the local plan with which it must conform. In the case of the WSHWNP, though, we were dealing with two parts to the same local plan, wherein the implications for our area in terms of the scale, nature and location of proposed development were markedly different in the emergent Part 2 plan compared to the adopted Part 1 plan. Part 1 of the local plan was entitled 'Strategic Sites and Policies' (VWHDC, 2016) and LPP2 was entitled 'Detailed Policies and Additional Sites' (VWHDC, 2018). In principle, if Part 1 is the strategic document, then Part 2 should not contain anything strategic, but the allocation of a Strategic Development Site at Dalton Barracks and Abingdon Airfield was specified in Part 2. Consequently, it could be argued that either the allocation in Part 2 is not strategic, or if it is strategic, it should be in Part 1. If it is not in fact strategic, then the neighbourhood plan would not be required to support it, and if it is strategic and should be in Part 1, then Part 1 of the local plan should be reviewed and updated to reflect this. However, challenging the relationship between the two parts of the local plan would achieve nothing in terms of the issue at stake for our community, which was not so much the allocation itself but the associated proposal to delete land from the Green Belt to deliver the proposed development.

Our communities were largely supportive of development at the Strategic Development Site as it was accepted that it was a sensible site for development, but they had strongly and consistently voiced their opposition to the removal of land from the Green Belt. Across the whole Designated Area, 62% of respondents to the questionnaire that had been delivered to every household indicated that they did not want any land to be removed from the Green Belt, a figure that rose to 84% in the village of Shippon (WSHWNPSG, 2018). While the latter figure might be expected given that Shippon is the settlement that will be most directly affected by the proposed development, 62% opposition to the removal of land from the Green Belt across the whole area is itself notable as most responses came from Wootton, which is already inset from the Green Belt and its residents might therefore be expected to care very little about whether a village in the other parish simply joins it in being outside the Green Belt. The fact that there was such solidarity across both parishes in opposition to the deletion of land from the Green Belt gave us a clear mandate to hold a firm line in our community-led neighbourhood plan, but this put us on a collision course with these 'strategic' elements of the local plan.

The starting point in our strategic handling of this situation was to distinguish between the strategic objective of the delivery of housing at the site to provide for the VWHDC's proportion of Oxford's unmet housing need and the presupposed need to delete land from the Green Belt to deliver that objective. Whereas the LPA considered both issues (housing, Green Belt) to be strategic, we considered the deletion of land from the Green Belt to be a means to an end rather than an end itself. The proposed change to the Green Belt, from our perspective, was not a strategic issue but a practical one, and therefore we could challenge it as our communities wanted us to. This distinction meant that we could simultaneously support the proposed development but argue against the deletion of land from the Green Belt, in line with community sentiments.

This is where the NPPF was helpful in formulating our position, as housing is explicitly identified as a national strategic priority and is therefore clearly strategic, but the Green Belt is specified in the NPPF in relation to the government's strong support for the Green Belt and the need to protect it. However, the NPPF makes allowance for changes to be made to the Green Belt so simply relegating changes to the Green Belt to a practical rather than strategic matter did not resolve the issue of the proposed changes to the Green Belt themselves. This is where stage two of our approach came into play as, if an LPA wishes to alter the boundaries of the Green Belt, paragraph 136 of the NPPF states that it must establish that there are exceptional circumstances for doing so, which need to be fully evidenced and justified (MHCLG, 2018a), and paragraph 138 further states that in such circumstances, consideration should first be given to previously developed land (PDL).

The grounds articulated in LPP2 as establishing the required exceptional circumstances included the fact that it was a large site, that the site was developable, that the site had unexpectedly become available, that it was close to Oxford and therefore well-placed in relation to the city that it was intended to serve, and that much or all of the site was brownfield (previously developed land) (VWHDC, 2018). In our representations to the consultation on LPP2, we challenged all these points, not in the sense that they were inaccurate but in that they did not establish the *need* to delete land from the Green Belt to develop the site. Size, developability, availability, proximity to Oxford and brownfield status in themselves – either individually or cumulatively – do not constitute exceptional circumstances as they do not establish the need for the site to be removed from the Green Belt in order to develop it. Moreover, the large proportion of the site that is brownfield mitigates against the need to remove the land from the Green Belt to develop it as the NPPF already allows for the development of brownfield land in the Green Belt without changing its designation as Green Belt. Although LPP2 did not specify the exact proportion of the site that is brownfield, phrases such as 'large areas' (paragraph 2.54), 'sizeable extent' (paragraph 2.56) and 'much of the site' (paragraph 2.73) cropped up frequently (VWHDC, 2018), and it was clear to me at the public examination of LPP2 that the entirety of the site was assumed to be brownfield.

Consequently, one of the explicit grounds upon which the local plan argued for the establishment of exceptional circumstances was the very grounds upon which we argued against this. Whereas the LPA argued that because the land was brownfield they could remove it from the Green Belt to develop it, we argued that because it was brownfield they did not need to remove it from the Green Belt to develop it. From our perspective, exceptional circumstances were not and could not be established for this site because the extent of brownfield land within the site meant that there was no need to delete land from the Green Belt to develop it, as required in Paragraphs 136 and 138 of the NPPF (MHCLG, 2018a).

Unsurprisingly, this generated considerable debate during the public examination of LPP2, demonstrating not only the contradictory positions held but the different evidence and arguments that are employed strategically to support different arguments. One argument for development effectively merged a spatially articulated justification for deleting that land from the Green Belt because it was brownfield with a temporal justification for deleting such a large area from the

Green Belt to accommodate development plans beyond the period of the local plan itself. It was argued that the local plan was seeking to secure longer-term provision for future development at the site, and that the site could accommodate up to 4,500 dwellings in the longer-term. The implication of this was that the removal of land from the Green Belt would be deemed necessary for this larger development even if it was not necessary for the immediate allocation of 1,200 dwellings in the local plan period (up to 2031). From our perspective, of course, this changed nothing: if the 1,200 dwellings could be accommodated on part of the site without deleting it from the Green Belt because it is brownfield, a significantly larger settlement could just as easily be accommodated on the whole site as it was all deemed to be brownfield.

A second argument suggested that in seeking to prevent the deletion of land from the Green Belt we were effectively constraining development, while a third argument proposed that in order to deliver a sustainable settlement at the site, it would need to be a sizeable settlement to provide for retail, schools and other facilities, public transport services, and so on, and would therefore require development of the whole site, for which it would need to be taken out of the Green Belt. Here I perceive an extreme version of the prioritisation of economic growth as the implicit assumption would seem to be that economic viability of the site is not sufficient: it is maximum profitability that is at stake. Although this introduced confusion to the examination as to whether the inspector was examining the proposal for 1,200 or 4,500 dwellings, the ultimate direction of attention to the 1,200 allocation was fully supported by the neighbourhood plan so we were very definitely not constraining development as relevant to the examination at hand. Further, as no objectively assessed housing need had been established for the site beyond 2031 and as the final capacity of the site had not yet been established, there was neither any allocation nor specified capacity to constrain. Indeed, we were attempting to inform that calculation of the site's capacity and future allocation of development in accordance with the needs of our communities, thereby seeking to contribute to the strategic determination of the meaningful capacity of the site with respect to all the vested interests around the table, not just the economic interests.

In any event, our core argument withstands these counterarguments: even if the whole site is required for development, if the whole site is brownfield it still does not need to be removed from the Green Belt to be developed. With this in mind, our position with respect to the Green Belt is clearly not simply a case of 'the Green Belt must stay' as a means of preventing any and all development but a case of 'the Green Belt staying does not impinge on the stated strategic objectives of the local plan', such that our neighbourhood plan objective for the Green Belt was entirely compatible with LPP2's strategic objective for housing.

The Green Belt remains vulnerable, though, as although brownfield land within the Green Belt can be developed, there are additional restrictions placed upon such development compared to non-Green Belt land, in order to protect the function of the site in terms of the purposes of the Green Belt (for example, preventing urban sprawl and maintaining openness). The desire on the part of those promoting the development to delete the land from the Green Belt was presumably predicated on the desire to remove these restrictions. It is precisely because of these restrictions that we were able to support the development providing the land remains within the Green Belt, as those restrictions would help to ensure that the development is appropriate for its setting, thereby protecting the place identity and place attachment upon which neighbourhood planning depends. It is the Green Belt that constrains development, and that is its purpose, but it would not constrain development to the extent that it would jeopardise the strategic objectives laid out in the local plan. To my mind, we had presented an alternative future for the Strategic Development Site that could accommodate all sides: the 1,200 dwellings could fulfil the necessary unmet need within the plan period on existing brownfield land in the Green Belt, a significantly larger (although as yet unquantified) settlement could be delivered beyond the plan period on existing brownfield land in the Green Belt, and the Green Belt could be left intact to protect the sense of place that it is intended to protect.

We had approached this matter strategically and on one level we had consensus in that all sides wanted to see the site developed to its fullest extent. The difference comes – as we saw in the previous chapter – in our divergent understandings of what constitutes 'its fullest extent'. While those advocating development understandably prioritise maximising the number of dwellings and profitability to meet housing targets and increase CIL contributions, for us the fullest extent should incorporate the aspirations and concerns of those living in and around the site, and the extent to which it can be developed in accordance with and as appropriate to its status as Green Belt. Such considerations should inform what is considered not only deliverable but appropriate for a site: this would bring neighbourhood planning properly to the table and allow for agonistic collaborative planning that recognises and utilises the capabilities of residents to think and act strategically and that accommodates all parties to some degree rather than some parties exclusively at the expense of the others. Being strategic and doing strategy are neither features of a particular spatial scale nor capabilities only possessed by professionals: thinking strategically is available to all and can be applied to any spatial scale. Any meaningful attempt to integrate the so-called strategic objectives of the local level and the presupposed detailed concerns of the neighbourhood level must move beyond enforced dualisms between artificially prescribed roles and capabilities to recognising and nurturing their mutualities and the potential that can be realised through doing so.

Strategy as spatial: site or area?

Although much of the attention locally has been on the proposals for the Strategic Development Site, the neighbourhood plan covers the whole of the Designated Area, and while as a steering group we wanted to ensure that we managed the development proposal as effectively as we could, we also wanted to do so – as far as possible – within a broader strategic approach for the whole of our area, rather than having a raft of separate policies for the proposed development. In direct contradiction to the regulations and guidance, we were daring to be strategic at the neighbourhood level and in so doing we unsettled the distinction not only between the local as strategic and the neighbourhood as not strategic, but also between being strategic in a generic spatial sense and being strategic in a temporally-oriented site-specific sense.

In this instance, we took our lead from the spatial strategy of the local plan, at least in so far as its settlement hierarchy is concerned. LPP1 described a settlement hierarchy which would be used to direct development proposals, such that development would be prioritised in market towns, which were considered most sustainable in terms of employment and transport connections, and towards the major development planned for the 'Science Vale' area (around Didcot, well outside our Designated Area and to the South of the district) (VWHDC, 2016). Further down the settlement hierarchy, and most relevant to our neighbourhood plan, villages are considered suitable for only limited infill while settlements not listed in the settlement hierarchy are deemed to be part of the open countryside and therefore not suitable for any development. Classified as a smaller village, Shippon should therefore be subjected to development only in the form of limited infill, while Whitecross is classified as open countryside and is therefore not deemed suitable for any development. However, the proposal for the Strategic Development Site was to merge the new development with both Shippon and Whitecross. Immediately, then, we were faced with questions as to both how the spatial and temporal types of strategy relate to each other and how the two parts of the local plan relate to each other. The spatial strategy (in Part 1) stipulated that development in Shippon and Whitecross should be limited or non-existent, but the temporal strategy for the new allocation proposed to superimpose a large new development onto these two settlements. Similarly, as the policies in Part 2 are specified as being detailed rather than strategic, then the settlement hierarchy and spatial strategy laid out in Part 1 should presumably still apply to Shippon and Whitecross, protecting them from being merged with the new development, but to my mind the Part 2 policies were being treated as if Part 1 did not exist.

A more practical question for us as a neighbourhood planning team was how to respond to this situation, and in this regard our spatial strategy was concerned less with the size of settlements than with the existence of distance or separation between settlements. This was one clear example of where we were very much in conformity with LPP1 (the version that had already been adopted), as this is the document that specified the settlement hierarchy and spatial strategy that we sought to reimpose on the proposals laid out in Part 2. We therefore sought to use the ambiguity in the relationship between Part 1 and Part 2 of the local plan to our own strategic ends by arguing that the Part 1 strategies supersede the non-strategic Part 2. Specifically, we highlighted in our representations to the public examination of LPP2 both the restrictions on development applied to our settlements by the settlement hierarchy and spatial strategy and other core (strategic) policies in LPP1 intended to protect the separation, character and distinctiveness of individual settlements.

Through both the public consultation on the neighbourhood plan and the Character Assessment that we commissioned (TDRC, 2018), the significance to the character of our area of the scattering of small, discrete, rural and historic settlements, and the value placed on this character by residents, became clear. The distinct identities and appearances of settlements, the accessibility and visibility of the rural environment in which they sit, and the personal identification on the part of residents with this character were frequently articulated in consultation responses. In light of this, a key aspiration for the neighbourhood plan was to protect these qualities as far as possible, and to that end, we developed three sets of protections: Strategic Vistas, Strategic Green Gaps and Local Green Spaces.

Strategic Vistas are views of or towards the countryside that are deemed significant to the character or quality of the settlements in our area and are mostly sweeping, panoramic views but also include smaller-scale, more intimate views of the rural environment where these are visually significant. They were identified through the Character Assessment (TDRC, 2018) and informed by the Green Belt Study (Hankinson Duckett Associates, 2017) which in part addressed visual character and openness.

Strategic Green Gaps specified areas between the outer limits of two or more settlements where the protection of the specified area is considered important to prevent the future coalescence of settlements or the reduction of the green space to such an extent that the discreteness and character of the settlements is put at risk. These were similarly identified through the Character Assessment (TDRC, 2018) and informed by the Green Belt Study (Hankinson Duckett Associates, 2017).

Local Green Spaces are green spaces that are deemed locally significant in some way, whether this is for visual or functional amenity, heritage reasons or environmental qualities such as tranquillity. They need to be demonstrably important to the community, proximate to the community that seeks to designate it and not extensive. They are usually accessible to the public and often are already in use as community facilities, such as playing fields or allotments, but public accessibility is not required for a space to be designated as a Local Green Space. As with the other designations, these were identified through the Character Assessment (TDRC, 2018) and through public consultation, especially via an online community mapping platform and at a public workshop designed to 'test the evidence' for policies for which a higher level of evidence is required. Examples in our neighbourhood plan include a recreation ground, playing fields, a green verge with an avenue of trees, and green or open spaces that are not in formal use.

There are two significant challenges concerning these designations. The first is that although Local Green Spaces are defined as formal designations that neighbourhood plans can establish, Strategic Vistas and Strategic Green Gaps are not similarly specified. There are, though, two countervailing factors that we hoped would support our position: previous neighbourhood plans have identified valuable 'glimpses' of rural views for protection in addition to more panoramic vistas, and the overarching strategic approach that we have adopted to the neighbourhood plan provides an 'in principle' basis for the designations that we seek to establish. Consequently, we had both precedence and the capacity to articulate our thinking in terms of principle on our side.

The second challenge to this strategic approach to settlement separation is specific to one Local Green Space designation, at the Dalton Barracks playing field in Shippon. This space was described in the Character Assessment as being important to the open feel and rural setting of Shippon, a village which it recognised as already suffering in terms of character from progressive urbanisation, increasing traffic and the stark contrast in style between the civilian and military parts of Shippon (TDRC, 2018). It was also identified through public consultation as not only important visually but also as a valuable resource for the military and civilian populations to come together for events such as open days. Although not open to the public, the public can book the site for their own use should they wish. Consequently, the neighbourhood plan sought to designate this site as a Local Green Space. However, the playing field was within the Strategic Development Site and was scheduled for development within phase one of the development (DIO, 2017) so was an especially contested proposal in our neighbourhood plan.

It is partly due to the potential for such clashes that the tiered approach to our spatial strategy perhaps carries special value, as the same plot of land might be covered by more than one designation, maximising the chances of protecting at least some of those gaps, spaces and vistas that are locally valued. It also highlights the lack of a clear distinction between a generic spatial principle and a site-specific policy, as generic spatial principles 'touch down' in specific places (sites) and site-specific policies have the potential to be articulated as generic spatial principles with any particularities covered in the exclusions, caveats and conditions that specify the application of the in-principle statement. This is most clearly evidenced in relation to the proposal within the Publication Draft of LPP2 that the new settlement be developed according to garden village principles (VWHDC, 2018), as our neighbourhood plan incorporated both in-principle statements that covered this site as much as the rest of our Designated Area and site-specific policies that acknowledged the import of this one allocation to our area and the specificities of the LPA's proposals for that development.

Garden villages are a recent development in the Garden City movement, an approach to urban planning introduced by Ebenezer Howard in 1898, whereby self-contained developments sought to combine the advantages of urban living with the benefits of a rural environment (Howard, 1902). In 2016, the government released a prospectus calling for applications from LPAs for funding to deliver a new suite of garden cities (DCLG, 2016), and a similar prospectus was updated and reissued in 2018 (MHCLG, 2018b). The Town and Country Planning Association (TCPA) defines a garden city as 'a holistically planned new settlement that enhances the natural environment and offers high-quality affordable housing and locally accessible work in beautiful, healthy and sociable communities' (TCPA, 2018: 3). The contemporary approach to garden cities seeks to put both sustainable development and wellbeing at the heart of planning for new settlements, and the TCPA has specified a set of principles underpinning such developments (TCPA, 2018). These include strong community engagement with community ownership and stewardship, a mix of housing and employment opportunities, combining the best of town and country while enhancing the natural environment, and providing strong recreational facilities and integrated transport systems which prioritise active and public modes of travel. In 2016, the DCLG publication *Locally-Led Garden Villages, Towns and Cities* had further specified that garden villages – distinct from garden cities – must be free-standing settlements: "The garden village must be a new discrete settlement, and not an extension of an existing town or village" (DCLG, 2016, paragraph 14).

In the case of the Strategic Development Site, the development is described in LPP2 in terms of garden village principles, yet it is also proposed to merge the new settlement with Shippon and therefore would not be a discrete settlement. It is partly based upon this contradiction that we argued against this merger of the two settlements in our consultation responses to LPP2. While it has been argued that more recent TCPA publications refer to garden suburbs or urban extensions, and while it is true that more recent publications do acknowledge diverse forms that garden-type developments might take (TCPA, 2017, 2018), this position overlooks two countervailing factors. The first is that these more recent publications themselves recognise and adopt the specification from the DCLG, 2016 document that garden villages should be stand-alone settlements, indicating that the different terms are neither synonymous in practice nor interchangeable in discourse. Indeed, one of these more recent TCPA publications explicitly quotes and upholds the definition from the 2016 prospectus of a garden village as a stand-alone settlement and not a bolted-on development (TCPA, 2017: 6). The second is that the proposed development could never be a suburb or urban extension of Shippon because Shippon – according to the LPA's own settlement hierarchy – is not a town but a village, in which only limited infill is considered appropriate under their own spatial strategy (VWHDC, 2016). As we have consistently argued, even the lower allocation of 1,200 dwellings cannot reasonably be considered infill, let alone the longer-term potential of up to 4,500 dwellings. It could be argued that the TCPA provides for flexibility in allowing the 'garden' label to be attached to a development that meets most of but not all the specified principles, but even this does not resolve the issue. I would argue that for the development to be described in terms of garden village principles, it must either integrate with Shippon as the village that it is and therefore be limited to the level of infill, or it must be delivered as its own entity and therefore be separated from Shippon. Alternatively, for it to be called an urban suburb or extension, it must be separated from Shippon as Shippon is not a town, and therefore rather than being an extension or suburb, it must be delivered as a stand-alone town. In any event, the only way in which I consider it possible for the proposed development to be delivered in garden or any other form that is consistent with the LPA's own settlement hierarchy and spatial strategy is as a stand-alone settlement.

This apparent contradiction between the spatial strategy in LPP1 and the development proposed in LPP2 further confuses the matter of what constitutes being strategic, paving the way for parish councils and neighbourhood planning teams to employ LPA strategies to their own strategic aims, which in turn leaves LPAs vulnerable to decisions being made through the public examination of their own local plan that go against their own interpretation. It is this very possibility that gave us hope despite both the system and the process being weighted heavily against us for the reasons outlined in the previous chapters.

While this suggests that LPAs do themselves no favours when they adopt a 'flexible' approach to the planning documents that they produce as it introduces opportunities for alternative perspectives to be aired and to become dominant, it also applies yet more pressure on neighbourhood planning teams, as there is no guarantee that their efforts to reinforce their preferred strategic interpretation will succeed, so they simply find themselves trying to cater for yet more possible outcomes. On the one hand, in our representations on LPP2, we highlighted the contradictions between the two parts of the local plan and used this to argue for the establishment of a buffer between Shippon and the new settlement to bring the proposals in LPP2 into line with the core policies in LPP1. On the other hand, we continued to develop the neighbourhood plan on the assumption that we would be unsuccessful in securing separation for Shippon through the consultation process for LPP2 and set about establishing our own buffer in the hope that we could get our plan made before the local plan was approved. Given the current lack of any clear green space between Shippon and the military site we could not immediately designate a Strategic Green Gap in this area, but several areas were identified as Local Green Spaces, which we worked up into a buffer to seek to deliver the separation sought by Shippon residents between their village and the new settlement. As the designation of such a buffer within a neighbourhood plan is not formally provided for in the regulations, we could not rely on this alone, so we retained our designation of Local Green Spaces that overlap with the buffer as a security measure.

In a similar way, but on a smaller scale, we also specified a buffer, which is also a Local Green Space, at the northern end of Whitecross (to the north of the development site), which sits alongside the Strategic Green Gap between Whitecross and the development site, as although most of the fields in between were removed from the proposed development by the time the Publication Draft of LPP2 was released, the top end of Whitecross was still proposed to abut the eastern boundary of the new settlement. As this contravenes our spatial strategy for the separation of settlements and the protection of their character and setting, we have applied multiple separation designations to make it clear that this area is strategically important in more than one way. We have therefore been strategic in the sense of generic spatial principles, and we have been strategic in the sense of being site-specific with a view to managing long-term impacts, and we have used each as a way of underpinning, reinforcing and compensating for the other in a thorough muddying of the waters between the two meanings of being strategic explored here.

Strategy as temporal: place or process?

The distinction drawn previously between two types of strategy implied that the targeted allocations and policies in LPP2 stipulated long-term intentions for development at specific sites, but it is entirely possible to be strategic in a temporal sense in relation to the process of neighbourhood planning as it is in relation to the place of the neighbourhood. This in turn will be shown to bring implications for what we consider the appropriate or most effective place of neighbourhood planning.

Early in August 2018, we submitted our neighbourhood plan to the LPA. Through our own consultation process, and reflecting debates in the literature (Parker, 2012; Sturzaker and Shaw, 2015), several parties had indicated that they thought we were acting prematurely in developing our plan before LPP2 had been adopted due to the need for neighbourhood plans to support the strategic aims of the relevant local plan. The spatial and temporal message about the relative positioning of local and neighbourhood plans was clear: a neighbourhood plan comes below and after a local plan. The logic behind this is articulated in early literature on neighbourhood planning which drew out the importance of a strategic planning context for the development of a neighbourhood plan, as is provided by a local plan (Parker, 2012). However, this rather presupposes clarity as to what is strategic and what is not, and again condemns neighbourhood planning to little more than sweeping up the crumbs. Such views not only overlook the fact that we did have a strategic context for our work in the form of LPP1 and the emerging LPP2, but also miss the point of the exercise for us entirely. Relegating neighbourhood plans to the lowest and last document in a planning hierarchy is one thing; preventing those responsible for their production from contributing to the production of a higher-level plan with which they are expected to conform is quite another. It closes off the greatest opportunities for neighbourhood concerns and aspirations to have the maximum possible impact given the tightly circumscribed arena of neighbourhood planning activity in the current system; opportunities that I consider neighbourhoods to have every right to pursue if they can. In fact, I would argue that they have a duty to do so where they have received a clear steer from their peers that indicates a desire to shift the trajectory of some aspect of that emergent local plan. We were obliged to conform with LPP1, which we did, and we were mindful of the emergent LPP2, yet the very fact that LPP2 was emergent provided opportunities for neighbourhood concerns and aspirations to be fed into the developmental work on LPP2.

While neighbourhood plans cannot challenge local plans once they are adopted, it was LPP1 that had been adopted, not LPP2, which was still under development. Once LPP2 is adopted, those aspects of LPP1 that LPP2 supersedes will be governed by LPP2 but the other aspects of LPP1 will remain in force. What happens with respect to the contradictions identified between the strategic policies in LPP1 and the strategic allocation in LPP2 remains unclear. The implications of this for the neighbourhood plan were that if our neighbourhood plan could be made before the Part 2 plan was adopted, we might be able to influence LPP2 with respect to issues such as the designation of green spaces and provision of buffers. If the local plan ended up being adopted before the neighbourhood plan was made those green spaces and buffers might become redundant as the strategic allocation at Dalton Barracks would override those designations. Effectively, the two plans were in a race to be completed first, although LPP2 had the benefit of added strategic weight as the higher-level plan.

On the one hand, then, it was crucial that we do everything possible to get our neighbourhood plan made before the local plan was adopted. On the other hand, the consultation requirements for local plans provided opportunities for the neighbourhood planning team, as well as the parish councils, to ensure that the views of residents were fed into consultation on the emerging LPP2. Through this consultation, we sought to influence the development of LPP2 so that – once adopted – it would be more consistent with what our communities wanted. Our engagement with LPP2 was an attempt to nudge the local plan into greater conformity with our emerging neighbourhood plan by the time it was adopted rather than simply waiting for LPP2 to be adopted without trying to shape its contents to suit our purposes. In this respect we in fact pursued two strategic avenues: 1) nudging the local plan into closer alignment with the aspirations and concerns of the neighbourhood, which entailed 2) racing to complete the neighbourhood plan as quickly as possible. As with our spatial strategy, this increased the chances of achieving some gains for the neighbourhood as the closer we crept to completing the neighbourhood plan, the stronger our position with respect to influencing the emerging local plan.

The time pressure that we were under to achieve this was immense. Not only is neighbourhood planning time-consuming and onerous in any circumstances, but the pressure that Oxfordshire local authorities were under to complete their local plans in accordance with the county's growth deal to secure extra funding to support the delivery of development infrastructure, applied additional pressure to us, too. With that part of the examination of LPP2 of relevance to us taking place before the summer recess, we had to accelerate our activity dramatically, all the while facing much less certainty than in many cases due to the radical change in development proposals for our area between LPP1 and LPP2, the extent and significance of disputes as to what constituted strategic issues, and the unfinished nature of LPP2. The parallel processes of development for the local plan and the neighbourhood plan thus provided both opportunities and challenges for the effectiveness of our neighbourhood planning efforts.

As detailed earlier in this chapter, we had adopted a strategic approach to both questioning the strategic assumptions of LPP2 and to the protection and maintenance of settlement separation in our area, but the determination to have as much influence as possible in shaping the local plan – and the acceleration of the process that entailed – were of crucial strategic importance. It meant that we had completed the Pre-Submission consultation before attending the examination and could therefore both draw upon the comments that supported our position and respond to those that did not. It meant that our contributions carried that bit more weight as we were that bit further down the line to completion. It meant that we had clear and firm ideas as to the preferred options for Strategic Vistas, Strategic Green Gaps, Local Green Spaces and buffers. It meant that we had a substantial and robust evidence base to support our proposals and contributions, and it meant that we knew where the areas of dis/agreement were between the local plan and the neighbourhood plan so that we could reinforce our position as supporting the strategic objectives but having specific concerns about key issues for our communities. Having adopted a strategic approach to settlement separation and character also meant that we could talk in terms of principle and strategy rather than being perceived as being NIMBYish in resisting development in specific places. All these factors enabled us to contribute to the public examination of LPP2 from a much stronger position than would otherwise have been the case, enhancing the possibility that we might be able to shape the final policy wording and development proposals enshrined in LPP2.

The more closely that policy wording in LPP2 reflects the desires and concerns of residents, the easier it would be to deliver a neighbourhood plan that is consistent with the objectives of the local plan, because those strategic objectives should entail fewer undesirable non-strategic elements, such as the merging of Shippon and the new development. This, though, raises an interesting scenario. If we were to achieve through our presentations at the public examination of LPP2 all the objectives that we had laid out in our neighbourhood plan, then we would not need our neighbourhood plan at all because the local plan would incorporate all those objectives. In that event, we would have spent thousands of hours producing a plan that would be redundant before it was even made, potentially rendering the whole neighbourhood planning exercise pointless. On the other hand, in this hypothetical scenario, we would only have been able to have such shaping influence on the contents of the local plan by virtue of our existence and activity as a neighbourhood planning team, and specifically, by virtue of having made sufficient progress on our neighbourhood plan to be able to contribute to the evolution of LPP2 *as if* the neighbourhood plan had been made even though it had not.

As has been articulated elsewhere, the question then arises as to whether it is the plan or the process that is most important (Bradley and Brownill, 2017). However, I think there is a broader issue at stake here, in that it is not a case of either/or, as we had to be sufficiently far along the process to be in the position of strength that we were in at the time of the public examination of LPP2. By that stage we did have a plan, albeit only a draft and far from made, but we had a formal plan document and all the supporting evidence to draw upon. The key lesson from our experience, I think, is that we need to move beyond talking about neighbourhood plans as being planning documents that sit below a local plan to talking about neighbourhood planning as a process that is intermeshed with the rest of the planning system, and that affords neighbourhood planning teams the chance to do more than simply produce an aspirational document for their area that follows meekly on from the relevant local plan but in reality achieves nothing of any substance. Neighbourhood planning teams do not just have the autonomy to develop a neighbourhood plan for their own area; they also have the autonomy to participate in broader planning processes and procedures, and to feed local views and concerns into those processes and procedures, with a view to shaping the very local plan with which the neighbourhood plan will have to conform. The comments of statutory bodies about our work being premature not only reflect a common presumption that a neighbourhood plan must be written after a local plan, but also I suspect some confusion among other stakeholders as to the relationship between LPP1 and LPP2 as they seem to think that we were operating in a strategic vacuum when in fact we were very much alert to the strategic threats that we faced and the strategic opportunities that were available to us.

Significantly, by contributing to the public examination of LPP2 as robustly as we could in part delivered on its potential even while the examination was ongoing, as the LPA was encouraged to reconsider aspects or issues that we had raised, and to liaise with us in relation to the definition of garden villages. While there are no guarantees that this will lead to better outcomes for our neighbourhood, it does at least start to unsettle the conventional positioning of a neighbourhood plan as coming after and below a local plan. Here, at least tentatively and potentially, a neighbourhood plan gained some traction in the examination process and might yet gain some influence on the local plan with which it must ultimately conform. It is therefore not just the place of the neighbourhood that can be addressed in a strategic manner, but the place of the neighbourhood plan within the broader planning system that can be reconfigured in a strategic manner. It might be unconventional, it might be frowned upon by other stakeholders, and it might be onerous and frustrating for those involved, but if it helps to nudge a local plan into closer alignment with the aspirations and concerns of a neighbourhood – however minimally – then perhaps it is worth it. Perhaps.

Given this situation, our engagement with the consultation and examination processes for LPP2 in our attempt to nudge LPP2 into closer alignment with neighbourhood priorities was at least as important as our work on the neighbourhood plan itself and – I would suggest – was crucial in securing whatever concessions (if any) we end up securing in LPP2, especially if the local plan is approved before the neighbourhood plan is made. The strategic approaches that we adopted to both defining the strategic in the first place and establishing a tiered framework to protect the separation and character of settlements were valuable in and of themselves, but they really came into their own in the context of the temporal strategy that we adopted to maximise our chances of shaping the content of the local plan with which we would have to conform. The temporal strategy enabled us to take a place at the table of the public examination of LPP2 *as if* the neighbourhood plan had been made, while our attempts to question or define the strategic and our overarching spatial strategy enabled us to speak from considered principles rather than specific details. It is therefore entirely within the capabilities of neighbourhoods to act strategically, in multiple ways, on multiple levels, and in relation to the place of the neighbourhood plan as much as to the place of the neighbourhood itself.

Conclusion

Contrary to assumptions that neighbourhood planning teams struggle with acting and articulating strategically, this chapter has explored the varied ways in which we adopted a strategic approach both to the development of our own neighbourhood plan and to how our plan might relate to the local plan with which it must conform. Feedback that we received advising us to remove reference to anything 'strategic' in our neighbourhood plan as only a local plan could establish strategy (WSHWNPSG, 2018) suggests that the very word 'strategic' has become anathema within neighbourhood planning, no matter what the intentions or capabilities of those involved might be. Moreover, criticism laid at the door of neighbourhood plans for being a NIMBY charter leading to a parochial obsession with inconsequential details and defensive resistance to any and all development (Parker, 2012; Sturzaker and Shaw, 2015; Bradley, 2017) is hugely disingenuous given the demands placed upon neighbourhood plans not to be strategic: if all that neighbourhood plans are deemed appropriate for is tidying up the details then it should come as no surprise that many such plans simply attend to details.

However, accepting such restrictions assumes a singular and narrow definition of what it is to be strategic, which is easily challenged by thinking through how strategy can be variously associated with scale, space and time. It is very easy for statutory bodies to extend the prohibition of neighbourhood plans from engaging in one very specific type of strategy to insist that they should engage in no form of strategic thinking at all, but even this one prohibited type of strategy becomes unstuck in light of the ability for a neighbourhood plan to allocate its own development sites, and to lay an absolute claim to strategic thinking at the level of the local plan is, I suggest, nonsensical. Our experience has shown that neighbourhood plans can be strategically developed in just the same way as a local plan; that strategy is not solely the territory of the LPA, nor is it aligned with just one spatial scale. Our experience has similarly confused the distinction between strategy in the sense of generic spatial principles and strategy in the sense of long-term intentions for one site, rendering any attempt to confine the strategic to the level of the local plan meaningless in both principle and practice. We have also stretched the idea of the strategic beyond the context of planning for the development of an area to apply it to the process of producing a development plan, raising questions as to the appropriate role and place of a neighbourhood plan in the broader planning system.

Evidently, in this instance at least, articulating neighbourhood aspirations and concerns in terms of strategy and principle is not the key challenge, nor is thinking and acting strategically a notable difficulty. Rather, the difficulties stem from the stipulation and presupposition that neighbourhoods will not, cannot, think and act strategically, and indeed from the assertion that neighbourhood plans should not even use the word strategic. With such restrictive preconceptions and the imposition of such linguistic constraints, is it any wonder that many neighbourhood plans do not appear to engage with strategy or articulate their aspirations in terms of principle? This prohibitive stipulation that strategy can only be applied in a monolithic fashion by a singular level within the planning system again leaves neighbourhoods with nothing better to do than determine inconsequential details and does them a significant disservice. As in the previous chapters, the planning deck is stacked against neighbourhoods despite claims to the contrary. The strategic capabilities and willingness of neighbourhood planning teams must be taken seriously and engaged meaningfully if neighbourhood planning is to survive in the long term and if it is to make any substantive contribution to planning, development and participatory citizenship. Chapter 6 deals with these prospects in more detail, but in the meantime, and in the current context of neighbourhood plans being simultaneously legitimised and delegitimised by the very process that breathed life into them, there is one more aspect of neighbourhood planning to be considered: if a neighbourhood's sense of place is not taken seriously, if their conceptual understandings and evaluative criteria are not taken seriously, and if their strategic capacities and perspectives are not taken seriously, what potential does the performative construction of place hold for the effectiveness of neighbourhood planning?

References

Bradley, Q (2017) Neighbourhood planning and the impact of place identity on housing development in England. *Planning Theory and Practice* 18 (2): 233–248

Bradley, Q and Brownill, S (2017) Reflections of neighbourhood planning: towards a progressive localism. In S Brownill and Q Bradley (eds) *Localism and neighbourhood planning: power to the people?* Policy Press: Bristol, p251–267

Brownill, S and Bradley, Q (2017) Introduction. In S Brownill and Q Bradley (eds) *Localism and neighbourhood planning: power to the people?* Policy Press: Bristol, p1–15

DCLG (2016) *Locally led garden villages, towns and cities* www.gov.uk/government/publications/locally-led-garden-villages-towns-and-cities

DIO (2017) *Dalton Barracks delivery document* www.whitehorsedc.gov.uk/java/support/dynamic_serve.jsp?ID=902932729&CODE=146712F47A0E0153F7A2FA74D0EA34AF Accessed 23 Jan 2019

Gallent, N (2014) Connecting to the citizenry? Supporting groups in community planning in England. In N Gallent and D Ciaffi (eds) *Community action and planning: contexts, drivers and outcomes*. Policy Press: Bristol, p301–322

Gallent, N and Robinson, S (2013) *Neighbourhood planning: communities, networks and governance*. Policy Press: Bristol

Hankinson Duckett Associates (2017) *Green Belt study – land surrounding Dalton Barracks* Available via LPP2 public examination library www.whitehorsedc.gov.uk/services-and-advice/planning-and-building/planning-policy/local-plan-2031-part-2

Howard, E (1902) *Garden cities of to-morrow*. Swan Sonnenschein and Co. Ltd: London

MHCLG (2018a) *National planning policy framework* www.gov.uk/government/publications/national-planning-policy-framework-2

MHCLG (2018b) *Garden communities: prospectus* www.gov.uk/government/publications/garden-communities-prospectus Accessed 22 Jan 2018

Parker, G (2012) *Neighbourhood planning: precursors, lessons and prospects*. 40th Joint Planning Law Conference, Oxford www.quadrilect.com/Gavin%20Parker.pdf. Accessed 15 Dec 2018

Parker, G (2014) Engaging neighbourhoods: experiences of transactive planning with communities in England. In N Gallent and D Ciaffi (eds) *Community action and planning: contexts, drivers and outcomes*. Policy Press: Bristol, p177–200

Parker, G, Lynn, T and Wargent, M (2015) Sticking to the script? The co-production of neighbourhood planning in England. *The Town Planning Review* 86 (5): 519–536

Parker, G, Salter, K and Wargent, M (2019) *Neighbourhood planning in practice*. Lund Humphries: London

Sturzaker, J and Shaw, D (2015) Localism in practice: lessons from a pioneer neighbourhood plan in England. *The Town Planning Review* 86 (5): 588–609

TCPA (2017) *Locating and consenting new garden cities: practical guide* www.tcpa.org.uk/guide-1-locating-and-consenting-new-garden-cities

TCPA (2018) *Understanding garden villages: an introductory guide* www.tcpa.org.uk/understanding-garden-villages

TDRC (2018) *Wootton and St Helen Without parishes character assessment* www.wshwnp.org.uk/wp-content/uploads/2018/08/WSHWNP-Character-Assessment-Introduction-full.pdf

VWHDC (2016) *Vale of white horse district council local plan 2031, part 1: strategic sites and policies* www.whitehorsedc.gov.uk/services-and-advice/planning-and-building/planning-policy/new-local-plan-2031-part-1-strategic-sites

VWHDC (2018) *Vale of white horse district council local plan 2031, part 2: detailed policies and additional sites* www.whitehorsedc.gov.uk/services-and-advice/planning-and-building/planning-policy/local-plan-2031-part-2

Wargent, M and Parker, G (2018) Re-imagining neighbourhood governance: the future of neighbourhood planning in England. *The Town Planning Review* 89 (4): 379–402

Wills, J (2016) *Locating localism: statecraft, citizenship and democracy*. Policy Press: Bristol

WSHWNPSG (2018) *Wootton and St Helen Without Neighbourhood Plan 2018–2031, consultation statement* www.whitehorsedc.gov.uk/sites/default/files/WSHWNP%20Consultation%20Statement_0.pdf

5 Performative constructions of place

Introduction

In this chapter, I attend to the post-political shifting of attention from substantive strategic issues to technical and localised details (Mouffe, 2005; Etherington and Jones, 2017; Kenis, 2018) to consider the varied ways in which neighbourhood planning as a process becomes increasingly de/re-politicised, effectively distracting those involved from the meaty issues of concern to their neighbourhood to the procedural obstacles that either arise as part and parcel of that process or that are introduced to the process by others, especially in circumstances where these can be used to obstruct the progress of a neighbourhood plan. Throughout the process, things will be said and done (or not) in a specific way by certain people in particular contexts and from the perspective of diverse agendas and divergent understandings; documents will be written in a similarly varied and contingent manner. Decisions will be taken (or not); consultees will (or not) provide their considered opinions; and actors will respond to events and correspondences in equally diverse and sometimes unexpected ways. The specifics of any one in/action might depend upon the immediate and longer-term context for the current situation, the aspirational goals being pursued, the anticipated responses of those with whom they are engaging, the stage they are at in the process, who might or might not have been advising them, how much time they have at their disposal to consider alternatives, and even their mood or the weather. In other words: specificities matter. However, these specificities take place in the context of huge inequalities in power and in a system that serves the interests of the already powerful (Bradley, 2018), so how much scope is there within neighbourhood planning for the relatively powerless to recraft what constitutes power? Which tactics and counter-tactics are employed, and how effective are they? This chapter, then, contributes further to addressing the identified need for greater attention to be paid to microdynamics of process to excavate new ways of bending the dialogic process to neighbourhood ends (Parker and Street, 2015; Parker et al, 2017; Vigar et al, 2017; Wargent and Parker, 2018).

My primary focus in this chapter is the crucial relationship between the neighbourhood planning team (including the respective parish councils) and the LPA, which is considered to be the central dynamic within neighbourhood planning (Parker et al, 2019). In our case, the relationship between the NPSG and the LPA in the early stages of plan development involved both occasional and welcome advice on the regulations and their implementation either by email or at NPSG meetings, and substantive, hands-on co-operation with respect to community engagement. We also benefitted from the local authority networks in sourcing data on social inclusion and other local characteristics, so the VWHDC was undeniably supportive of our neighbourhood planning process at the stage of intelligence and evidence gathering. As the neighbourhood plan started to take shape as a document, though, the different agendas in relation to the Green Belt, garden village principles and settlement separation became clearer and these cordial relations became increasingly strained. The different ways in which this relationship was performed form the focus of this chapter.

In the first section, I explore the scope for LPAs to invoke formal political hierarchies that had lain dormant to undermine the autonomy of the neighbourhood planning group, potentially modifying the very process of neighbourhood planning itself (Parker et al, 2017) and remaking (unmaking) the place of the neighbourhood. Subsequently, I direct attention to a softer form of power available to LPAs, performed under the rhetoric of helping the neighbourhood planning team, with its practical emphasis on getting things done (Parker et al, 2015). Finally, the difficulties in balancing the requirements of multiple audiences on the part of both neighbourhood plan and local plan are explored, revisiting the issue of different knowledges and languages but in the context of the oral as much as the written presentation of neighbourhood concerns and aspirations. In all three cases, an LPA is in a powerful position to identify (introduce?) potential hurdles and to present themselves as the procedural saviour of a neighbourhood plan. This not only has the potential to impede the progress of a neighbourhood plan significantly and divert attention away from substantive aims to technical details, but also potentially enforces the dependency of the neighbourhood plan on the LPA. I argue that such a move could in extreme cases constitute abuse of power as it would extend beyond simple institutional oversight (Parker et al, 2017), potentially upending the very ethos of neighbourhood planning and repoliticising the process – and especially its burdensome nature – in a highly troubling manner.

Political constructions of place

Neighbourhood planning is undeniably a political process. It is bound into political structures and procedures and it is all about power dynamics. In our case, the neighbourhood plan was initiated by two parish councils, and although the steering group is predominantly populated by non-councillors, it also includes representatives of each parish council. Finding ourselves in a position where our communities were adamant that the Green Belt must remain Green Belt, but unable to take an oppositional stance to the strategic objectives of the local plan, we needed to find a way to protect the Green Belt in a constructive fashion.

In the first instance, we sought to draw apart the two issues that the local plan seemingly presupposed were one-and-the-same thing: the strategic allocation and the Green Belt. This generated our argument that deleting land from the Green

Belt was not strategic but practical. Secondly, we needed to find a way of framing the Green Belt as something other than 'that which obstructs development', for which the Character Assessment (TDRC, 2018) that we commissioned was helpful in highlighting the historic lineage of the rural and specifically agricultural character of the Designated Area. This enabled us to evidence that the Green Belt is not merely a nicety that constrains development but a fundamental recognition and protection of the historic rural character of our area, shifting the terms of the debate away from accusations of NIMBYism by establishing the Green Belt as a matter of principle. The argumentative substance of our position, however, was only part of the issue, as our temporal strategy prompted the activation of explicitly political tactics.

In the previous chapter, I outlined our strategic approach to try to drive the neighbourhood plan through before LPP2 could be approved. While this undoubtedly made life more difficult for us it probably also had the same impact on the local authority, whose priority – understandably – was the progress of LPP2. Once it was confirmed that we intended to press on with the neighbourhood plan in parallel to the local plan, it seemed to me that the LPA took steps to take control of the process. The LPA met with representatives of the WSHWNP Steering Group and then met separately with parish council representatives on the same day and, at another meeting later that day between members of the steering group and the parish councils, it seemed that the LPA's articulated position and perspective had been markedly different between the two meetings that morning. Those who had attended the first meeting reported that the LPA had commended them for the progress that they had made and the consultation that had been undertaken and had offered further support for the refinement of the emerging policies. Those attending the second meeting, by contrast, reported that they had been warned that it would be premature to press on with the neighbourhood plan while LPP2 was still under development, that the Pre-Submission consultation process would need to be re-run, and that many of our policies would inevitably be deleted at examination.

I admit to perceiving this as a classic attempt to divide and conquer, separating the NPSG from their parish council sponsors and applying political pressure through the formal local government structure that pre-existed neighbourhood planning. However, I also consider it a spatial reconstitution of the Designated Area, dismantling it back into its two constituent parishes thereby undermining horizontal unity as well as vertical unity. Given the time pressure resulting from aligning ourselves with the timescale for LPP2, this purported risk of needing to re-run the statutory six-weeks' consultation understandably caused concern, but all parties remained firm in their commitment to progress the neighbourhood plan in accordance with our temporal strategy. I consider this perceived attempt to derail the neighbourhood plan to have boosted our resolve as we took it to show that we had more power at our disposal then we had previously realised. Incidentally, shortly after that day of meetings, we received notification that a planned policy workshop with the LPA would no longer take place. While I cannot and do not comment on the motivations of the LPA in this regard, this does highlight the potential for volatility and conditionality in local authority support for neighbourhood planning. This in turn renders the provision of support amenable to being utilised as a political weapon by local authorities that are so inclined, drawing attention to the lack of clarity identified in the literature as to what constitutes an LPA's duty to support neighbourhood planning (Parker et al, 2015, 2017). Without such clarity and in extreme cases, neighbourhood planning teams are conceivably vulnerable to fickle and politically motivated 'support' that potentially undermines the whole purpose of neighbourhood planning, let alone the objectives of any specific neighbourhood plan.

The highly and unavoidably political nature of the neighbourhood planning process, then, had implications for the speed at and attitude with which the neighbourhood plan was pursued, and prompted certain responses and in/actions on the part of both parties. The actions of the local authority – which we perceived to be for certain purposes – generated effects that I suspect the authority would not have anticipated, which in turn stimulated further responses on their part. Perhaps reflecting the formal role of local authorities as knowledgeable guides and overseers in neighbourhood planning, it seemed to me that we were expected to take the advice of the LPA without question and that to do anything else would be deemed obstructive, confrontational and irresponsible irrespective of our different political positionings and objectives. The perceived implication of irresponsibility is significant here, as it seemingly suggests preconceptions of local residents as incapable of either understanding the issues and processes involved in planning or making sensible decisions. Conceivably, such preconceptions might have led to the perceived approach made to parish councillors who are presumably deemed better equipped to make appropriate decisions and could therefore be brought to bear on the NPSG to put it in its rightful place at the back of the planning queue.

This again portrays a perception that there is a very particular 'appropriate place' for and of neighbourhood planning within the structures of planning. It comes after local planning; it comes below local planning, even for a local plan that has not yet been adopted; and it can be community generated until it needs to be contained, at which point it can be brought back into the fold of formal political structures. As discussed in the previous chapter, though, we disrupted all three of these assumptions of neighbourhood planning. We sought to pre-empt an emerging local plan, and thereby positioned ourselves as at least parallel to that emerging plan in terms of both position and timing, and even when challenged by the perceived imposition of authority within the formal political structure, we remained community-led because of the strong mutual support between the steering group and the parish councils. Consequently, not only does neighbourhood planning revolve around the politics of place-making but the place of neighbourhood planning can be highly politicised and can be challenged through the very process of neighbourhood planning. While this might be argued to demonstrate a re-politicisation and the rise of a new antagonism in neighbourhood planning in contrast to the concerns over a post-political denial of 'politics proper' (Bradley, 2015), this would – I think – overstate the case as the type of politicking discussed here does not revolve around substantive concerns over the model or scale of development but around technical and administrative concerns as to who has the power to determine what is written where.

This perspective is reinforced by events that unfolded later in the planning process, after the public examination of LPP2 but before that of the neighbourhood plan, which raise again the different views held within these documents regarding what constitutes garden village principles. The consultation draft of LPP2 described the new settlement as reflecting garden village principles even though they planned to merge it with the village of Shippon. By contrast, we argued that if it is a garden village then it must be separated from Shippon as a discrete settlement in accordance with the TCPA guidance (TCPA, 2017, 2018) and if it is a garden town or urban extension then it must also be separated from Shippon because Shippon is a village and not a town according to the settlement hierarchy and spatial strategy in LPP1 (VWHDC, 2016). Representatives of the NPSG and parish councils received notification of extraordinary meetings to be held by the LPA to approve the proposed amendments to LPP2 following examination. Neighbourhood representatives attended these meetings and discovered through them that the LPA had applied for government funding (MHCLG, 2018) to support the development of a 'garden community' at Dalton Barracks and Abingdon Airfield. The text of this application is highly illuminating with respect to the relationship between the LPA and the neighbourhood planning team, as it articulated WSHWNP support for the VWHDC's garden village proposals without making any mention of the marked differences in opinion as to what constitutes garden village principles, and it applied for funding for a transformational garden community that can be superimposed onto an existing settlement rather than a new, discrete settlement despite community support for the latter but not the former. The application also stated that the WSHWNP team had been consulted about the application and that the LPA was awaiting a response from us, but as we had no knowledge of this application until the extraordinary meetings we would disagree. We could easily (and I do) perceive this as a deliberate exclusion of the neighbourhood planning team from this process but the main point here is that – whether by oversight or intent – the concerns and aspirations of the neighbourhood as clearly, consistently and robustly articulated both in the neighbourhood plan itself and through our involvement with LPP2 were not reflected in our reading of the application sent to MHCLG. To my mind, this not only flies in the face of the spirit and intentions of the Localism Act but also makes a mockery of the processes for both neighbourhood planning on its own terms and for public engagement in local plans. It also reignited the explicitly political aspect of neighbourhood planning as it prompted us to consider and decide how to make our proper views known. We wrote to the MHCLG via the website for the Garden Communities Prospectus outlining our grave concerns about the LPA's application, requesting that our letter be forwarded to the evaluation panel assessing the applications.

As yet, we have no idea whether this will be done or whether it will have any effect, but our approach has been consistent and our determination resolute: if we feel that we are denied a voice or feel excluded from a process in which we should rightfully have a place, we will seek alternative means by which to reinsert ourselves into the process because we are committed to doing whatever we can in the interests of our neighbourhood. This action was not designed or intended to make life more difficult for the LPA – although it will no doubt be perceived that way by some – but if we perceive that we are being denied the opportunity to engage with planning in the manner that was intended by those who are duty bound to support our activity, then we will create other ways to engage and if that introduces difficulties for those who prompted the need for such creativity in the first place, so be it. Legitimate engagement is always, surely, preferable to alternative forms of engagement but it does at times seem to me as though we are the only ones willing to engage legitimately.

Neighbourhood planning, then, is political through and through, and political structures and processes can be used by those further up the planning system to try to control the progress of a neighbourhood plan and the achievement of its objectives, but these attempts are not necessarily effective and can lead to unexpected twists in the neighbourhood planning process. At the very least, our experience shows that neighbourhood planning teams can set and sustain their own focus and timescale even in the face of political challenge, and that neighbourhood planning can take place in a host of spaces other than the neighbourhood plan itself. Whether by interweaving with the local plan, reconfiguring the spatiality of the planning hierarchy entirely, or by working around the formal planning system to access national level decision-makers, neighbourhood planning has the potential to relocate itself away from the position of coming after and below a local plan, and can reposition itself to do far more than simply sweep up the crumbs and fret over inconsequential detail. Our neighbourhood plan is very definitely community-led both in content and in timescale: we were clear in our objective to do whatever we could to protect the Green Belt (and the separation between settlements that it helps to secure) in accordance with community interests and we were clear that our best chance of doing that meant progressing our neighbourhood plan as far as possible before the local plan could be approved to influence the local plan so that it more closely reflected our own ambitions. We have also become increasingly clear through these political manoeuvrings that we will resist any perceived attempt to deny, silence or ignore our legitimate perspective to the best of our ability. Whatever the outcome, these political elements of our neighbourhood planning process will have played a specific role in determining its final effectiveness in delivering the objectives of the community, and therefore in determining both the geographical place to which it relates and the structural-procedural space in which it takes place.

Practical constructions of place

As well as the potential wielding of formal power by LPAs as a means of trying to control the direction and progress of a neighbourhood plan, there exist softer ways in which those in authority can either shape neighbourhood planning policies or manipulate the neighbourhood planning process, thereby subsuming or re-scripting community aspirations (Parker et al, 2015, 2019; Vigar et al, 2017), nominally under the aegis of providing procedural 'help' to the neighbourhood planning team. The advice provided to us was often framed as helping to ensure that we did not fall foul of duplicating national or local policies, including non-planning issues, drafting policies that could not be applied, or otherwise tripping ourselves up at examination. I am sure that many of these offers of assistance were genuine, but the emphasis did seem to me to slide towards ensuring that the neighbourhood plan would not interfere with the finer practicalities and details of the emerging LPP2 rather than ensuring that we were in broad conformity with its strategic elements or supporting us in delivering community objectives as the process reached its closing stages. We saw in the previous chapter that what counts as strategic is itself up for debate, and the evaluation of 'broad conformity' is equally under-specified in the regulations (Wargent and Parker, 2018), so when these two coincide there is clearly much scope for confusion and conflict.

Political modulation of community objectives has been identified as taking two forms: the exertion of a modifying or controlling influence by those in authority over the community or neighbourhood, and the modulation of strength, tone or pitch of voice by the community or neighbourhood in response (Parker and Street, 2015). On this reading, political modulation is done to community representatives by those in power through a host of technical tactics, such as deflecting responsibility, withholding information, speeding up or slowing down the timetable and localising issues, while those community representatives simply raise, lower or embolden their voice as a means of defence or defiance. While I can perceive many of the preceding examples of modulation in my own experience of neighbourhood planning, I would question the unidirectionality of influence that is assumed here. For example, it seemed to me that a separate document being developed to guide work in relation to the Strategic Development Site was deemed not to be part of the local plan for the purposes of our own concerns but was deemed part of the local plan for the purposes of our obligation to conform with it. I also consider that our progress was delayed by the arrangement of successive meetings at which LPA concerns over policy wording were due to be resolved, which became intertwined with inevitable difficulties in arranging meetings at all due to the pressures that LPP2 placed on the LPA. One such meeting was described as informal, in light of which the LPA did not provide us with an agenda or any indication as to which policies were at stake or what their concerns related to. We were then informed that despite this being billed as an informal meeting, the LPA's legal counsel would be in attendance, which to my mind placed it firmly in the 'anything *but* informal' category. The combination that I perceived of the LPA hiding behind a legal barricade that we might be expected to find daunting and obfuscating as to the content of the meeting to prevent us from preparing for that meeting was admittedly unsettling, but it also enhanced the sense that if the LPA felt it necessary to resort to such measures, then perhaps we had a stronger hand than we realised. Thankfully, we were able to commission expert support for this meeting, thereby establishing more of a balance between the parties than would otherwise have been the case and going some way towards counteracting what I personally consider to be a clear abuse of power on the part of the LPA.

I certainly would not deny that there are many and varied ways in which LPAs can modulate the objectives of a neighbourhood plan and/or manipulate the process, both legitimate and otherwise, nor would I deny that modulating one's voice is an option available to residents in response: I certainly felt (and was probably perceived to be) increasingly belligerent as the process rolled on. However, as outlined in the previous section, there is a far greater range of options available to community representatives than simply modulating their voice. While LPAs might seek to deflect an issue, there are ways in which neighbourhoods can keep that issue on the table, for example by questioning what it means to be strategic, thereby rendering uncertain LPA efforts to place specific issues out of bounds. Similarly, while LPAs might seek to vary the speed of the neighbourhood planning process, neighbourhoods are not always powerless in determining whether to allow this to happen, and in some circumstances can accelerate their own process, effectively putting themselves in charge of the process – at least for a while – and the LPA on the back foot. The well-documented reliance of neighbourhood planning teams on external professional expertise to accommodate imbalances in power, skills and knowledge also plays a role here (Gallent and Robinson, 2013; Bradley and Brownill, 2017; Bradley, 2018; Brownill and Bradley, 2017; Parker, 2017; Wargent and Parker, 2018), as tactics unavailable to first-time community volunteers are more readily available to trained and experienced planning professionals who know how other trained and experienced planning professionals operate. In this context, the reliance on such expertise can be an absolute necessity, and at least as valuable as their contribution in articulating neighbourhood ambitions in the language of planning. The power games that can be played during the formalisation of neighbourhood aspirations and concerns are far more significant to my mind than the precise wording of policies: wording can be debated later; the inclusion or exclusion of major concerns underpins the very possibility or otherwise of even entertaining those wording debates.

Perhaps the most pervasive and powerful way in which I consider that the LPA – in practice even if not by design – potentially modulated our neighbourhood planning process, though, came not from anything explicitly articulated but a sense that I acquired that LPP2 was being treated as if it had already been adopted. While officially it was acknowledged that the process was incomplete, it seemed to be implicitly assumed that its draft content would go forward without major modification and we therefore had no choice but to bring our neighbourhood plan into line with the draft LPP2. Associated with this was an apparent assumption that LPP1 was already null and void, even though it would continue to apply beyond the adoption of LPP2 except where explicitly superseded by LPP2. Meanwhile, it seemed to me that we were repeatedly reminded that our own neighbourhood plan was a long way off being made, that it would probably be significantly modified at examination, and that therefore it was deemed to carry no weight at all, even though the closer it crept to being made the more weight it acquired.

Such perceived manoeuvrings might be expected in a political scenario in which two parties disagree on fundamental matters, but my bigger concern here is the normative emphasis on helpfulness and its association with procedural requirements. The arduous and time-consuming nature of neighbourhood planning, its high demands in terms of local government, planning and technical expertise, and the significance of reaching procedural milestones in motivating ongoing activity have all been recognised in the literature (Gunn et al, 2014; Parker et al, 2015; Vigar et al, 2017; Wargent and Parker, 2018) and I would not challenge any of those observations. My concern, though, is that this burdensome nature of neighbourhood planning is itself open to use and abuse for political ends. The pressure that can be brought to bear on neighbourhood planning teams to modify the wording of their policies, to delete entire policies or to add or remove conditions and caveats to/from policies in order for the neighbourhood plan to progress to the next stage of the process is significant, as it can be claimed that the plan would not meet the Basic Conditions requirements, or the Strategic Environmental Assessment requirements, or would end up being deleted at examination for any of an unspecified number of reasons. These might, of course, be genuine dangers, but they equally might not and neighbourhood planning teams, by and large, are not going to be sufficiently equipped to tell the difference, especially if they do not have sufficient resources to buy in expert advice on such matters.

One thing this highlights is that if an LPA does want to use this burdensomeness for political purposes by heaping more and more work and pressure onto an NPSG to force their capitulation to see their plan progress to the next stage, then the LPA had better be confident that this burdensomeness is the greatest concern for the neighbourhood planning team. I am not for one moment suggesting here that this was the intention of the VWHDC, but our example serves nicely as a hypothetical scenario. As indicated in the previous section, any perception on our part that the LPA might be trying to force our hand in a certain direction was more than likely going to force it in the opposite direction because I (we?) resented the sense that we were simply pawns in someone else's game to be pushed around at will. Beyond that, though, far more important to us than the ease with which we progressed through the process was – I think – the effectiveness with which we delivered on the ambitions of the neighbourhood. Yes, we wanted to reach milestones and yes, the ideal would be to work collaboratively with the LPA to reach completion (Parker et al, 2015, 2019) but milestones are meaningless and completion is pointless if the substantive content to which the process relates has been hollowed out. Assertions that we would end up having policies deleted at examination therefore had little effect on us, because the examination was another opportunity for us to present our perspective. Although resolving issues at examination took the decision out of our own hands it – more importantly – took the decision out of the hands of the LPA. It sustained the possibility that the decision, although not ours, might go our way, but it also highlights just how little power has in fact been transferred to communities under the Localism Act, as many community outcomes are now dependent not solely upon the LPA but on national level powers.

The fact that we ended up negotiating wording for so few policies, though, does suggest that as a strategic document (which it is not meant to be), our neighbourhood plan barely touches the edges of what the local plan was trying to achieve. On the one hand, this reflects genuine overlap in objectives in that each side wanted to see development at the Strategic Development Site, but on the other hand it speaks to the massively constrained arena for influence granted to neighbourhood plans if even our relatively brazen efforts barely scratched the surface. The space of neighbourhood planning is miniscule and is miniaturised even further if LPAs attempt to contain it through practical assistance which is open to abuse for the purposes of the LPA rather than in the service of the community. This space must be made more expansive if the statutory status afforded to neighbourhood plans is to have any meaning.

Presentational constructions of place

A neighbourhood plan itself, of course, is a presentational construction of place but in the process of developing a neighbourhood plan, presentational constructions of place can be considered in relation to other aspects, too, including the varied audiences of a neighbourhood plan and the presentation of the neighbourhood planning team (and thereby the plan itself) in broader planning contexts.

As a document, a neighbourhood plan is very much stuck in limbo between two worlds. Once made, it is an official planning document with legal weight and as such its primary audience is the LPA that will have responsibility for using the document to help determine planning applications that affect its Designated Area. However, as a community-led plan, the primary audience for a neighbourhood plan is the community whose (planning-related) views, concerns and aspirations are encapsulated in the policies therein. If a neighbourhood plan also identifies community projects that are taken forward by the parish council or a different form of implementation group, that community-oriented side of the neighbourhood plan will be magnified. While the policies must be operable in a legal sense to enable the plan to perform its formal planning function, it must also be intelligible to the local people whose concerns it is intended to address. This Janus-faced nature of a neighbourhood plan not only reflects the central tension in the whole neighbourhood planning process between lay knowledge and sentiments and official discourse and protocol, but also complicates the process of writing the plan which in turn has implications for the future of the place being planned.

While the policies themselves are required to be written in the formal language of planning, the supporting text is not. It is in this supporting text that strength of local feeling about, evidence gathered in relation to, and justification for, the specific policies are normally presented and explained. However, such explanatory text can also serve other purposes, such as:

1 Demonstrating to the community that we have taken their views on board even where it is not possible to include a specific policy to address those views (for example, because it is not planning-based or because there is insufficient evidence).

2 Representing the strength of local feeling on issues that might conflict with the local plan or that might duplicate national or local policies (such as the Green Belt, our support for the protection of which both duplicates the NPPF and conflicts with the LPA's stated intention to delete land from the Green Belt).
3 Explaining how the neighbourhood plan relates to the local plan and what opportunities it does and does not bring with it.
4 Informing residents of potential development and associated impacts in our area.
5 Garnering interest in and support for the aims and process of neighbourhood planning in our area.
6 Reassuring residents that the plan is genuinely community-led and is not open to abuse by the local authority, in terms of either the development agenda for the area or the data and views provided by residents as part of the consultation exercise.

However, the role of the LPA comes into effect again here, as their focus on the applicability of policies potentially blinds them to the supporting, explanatory text which can be important to the interpretation of a policy and the intent behind it (Parker et al, 2019). During our negotiations about policy wording we were advised that we might need to conduct an SEA because it looked as though one of our policies allocated development to the Strategic Development Site, despite the fact that our plan (in the supporting text) states on multiple occasions that given the size of the planned development in LPP2 we do not allocate any additional sites or development (WSHWNPSG, 2018b). This was also made clear in our SEA Screening Opinion, which found that no SEA was required explicitly and specifically because we were not allocating any development (WSHWNPSG, 2018c). This charge of needing to conduct an SEA could be (and was) perceived as yet another attempt to stall or obstruct our progress, but regardless of the motivation, the key issue that it raises is the disconnect between the policies and the supporting text and the way in which this holds apart the neighbourhood as a planning space and the neighbourhood as a community place. Embodying two types of knowledge in one document does not in itself constitute bringing local knowledge into planning if planning concerns itself only with that part of the document that speaks its own language. Once again, the ambitions of neighbourhood planning fail to materialise.

Explaining how the neighbourhood plan relates to the local plan is also important as consultation feedback revealed that many residents are unclear about this. This becomes especially important in the later stages of the process as the local authority arranges both the public examination and the referendum, thereby distancing the neighbourhood plan from those responsible for developing it. No matter how community-led the policies might be and no matter how robustly we maintained a community focus in relation to both the plan's content and timeframe for production, the final communications about it will come from the LPA and not the neighbourhood planning team or the parish councils. This is a significant source of confusion, which threatens to undermine the perceived independence of a neighbourhood plan even if a relatively high level of independence has been maintained throughout the process, with implications for both community and volunteer engagement. If residents consider the neighbourhood plan to be just another planning document churned out by 'the council' which they expect to do them no favours whatsoever based on previous experience or anti-council bias, how realistic is it to expect them to vote in a referendum? Similarly, why would volunteers remain involved in neighbourhood planning if their best efforts to serve their neighbourhoods and address some of the perceived shortcomings in LPA attitudes and behaviours are indiscernible because the very process with which they have engaged masks their effort, positionality and intention? Such issues must be dealt with if neighbourhood planning is to realise its potential in relation to both what it can do and what it can be seen to do. It also further muddies the waters with respect to the place and role of neighbourhood planning in the wider planning system. Simultaneously, neighbourhood plans are kept at arms' length from local planning in that they are not permitted to do what is deemed to be a local plan function (strategy) or to duplicate local plan policies (no matter how keenly felt at the neighbourhood level) while also being 'under the wing' of the local authority in terms of arranging the examination and referendum and 'under the thumb' of the LPA in terms of the sign-off required at various stages (e.g. setting the Designated Area and progressing to examination). This lack of clarity and inherent conflict in terms of role, relationship and ownership is a millstone around the neck of neighbourhood planning, which can only be partially addressed by specifying the duty to support or establishing memoranda of understanding (Parker et al, 2015, 2017).

While the formal requirements might be off-putting for residents, the need to be public facing has the potential to cause difficulties for the local authority, if it feels threatened or undermined by statements in a neighbourhood plan that are intended for public rather than local government ears. Examples in our case included criticism that we were not sufficiently supportive of the LPA's strategic objectives, and questions as to why we had stated that the questionnaire data had not been made available to either the district or the county council (WSHWNPSG, 2018a). In the former case, we stated in the foreword that we supported the development but not the deletion of land from the Green Belt and that while we supported the ambition to develop the site as a garden village, we expected garden village principles to be applied to the development in full and not in part in line with community sentiments (WSHWNPSG, 2018b). In terms of the questionnaire data not being made available to the district and county council, this statement was explicitly intended to reassure residents about our handling of their personal data in the questionnaire responses as we had received comments suggesting that they would simply be used by 'the councils' for their own agendas. While the concerns that were raised by the LPA at this level of detail somewhat contradicts the view stated earlier that the focus on policies might lead to neglect of the supporting text, the key point here is not so much the detail but my perception that the LPA seemed either unable or unwilling to read the draft neighbourhood plan from any perspective other than their own vested interest in proposing the development. It seemed to me that the LPA did not consider that we might be speaking to more than one audience, or that they perhaps ought to be wearing more than one hat when reviewing a draft neighbourhood plan: a local authority does not function only as a strategic driver for development but as a democratic body representing the needs of all its residents. The Janus-faced nature of neighbourhood planning, then, seems to present problems for both the neighbourhood planning team and the LPA, as well as the public.

When it came to representing the neighbourhood plan at the public examination of LPP2, though, that Janus-faced nature was far less in evidence than in the drafting of the neighbourhood plan because although the examination process was open to the public, there were very few members of the public in attendance and those that were in attendance were relegated to watch proceedings on a screen in an ante-room in a telling if unintended performance of the chasm between professional planning and local people. The examination process itself was formal, formulaic and highly orchestrated, with strict but largely unstated protocols for behaviour, such that anyone new to the process very much had to learn through doing. Unsurprisingly, this put the neighbourhood plan and parish council representatives at a distinct disadvantage as the district council, planners, site promoters and other agencies largely had previous experience of precisely this form of engagement whereas we did not.

The key to success – or at least acceptability – in this environment seems to be to speak in the right kind of language: technical, expert, professional. As a neighbourhood planning team we were potentially at a disadvantage on all three fronts: we had the wrong kind of knowledge (in that our knowledge was local, lay and non-scientific), we lacked the right language (in that we were not planners and had no legal expertise), and we had no power (in that we were non-professional volunteers with an as-yet-unmade neighbourhood plan). Consequently, when neighbourhood planning teams interact with the broader planning process, the supposed accommodation of the non-professional under localism is even less in evidence than in neighbourhood planning itself, suggesting that localism is not a new principle of planning but is itself localised to the lowest level of planning. Nevertheless, if we can learn to present the wrong type of knowledge in the right kind of language, then perhaps we can obtain a little more power than would otherwise be the case. When I talk about language here, I do not mean just the spoken word but the manner of engagement with the different perspectives around the table and the balance between appropriately articulated support for those aspects that must be supported as a counterfoil to those elements of a local plan that it is most important to challenge.

It is also about being prudent and selective about which battles are worth fighting in this kind of arena, as any one person's contribution to the debate is not determined by that person, but by the inspector who is chairing the examination. It is therefore very much about prioritising, which for us meant keeping our emphasis on the Green Belt, settlement separation and garden village principles, with secondary narratives around transport infrastructure, social (in)justice and environmental protection. Unsettling the conventional place of a neighbourhood plan as coming after and below a local plan, then, required us to engage in the broader planning system as if we were not neighbourhood planning volunteers and as if we were planning practitioners: if we were to be taken seriously, we were expected to have the requisite technical knowledge, speak in the appropriate language, and to do so with authority and assuredness. One important aspect of this was to speak in terms of principle, drawing directly upon that strategic aspect of our neighbourhood plan that was deemed inappropriate, and it did seem that the 'in principle' or strategic nature of our approach had been acknowledged, at least in relation to discussion about a site that we wanted to designate as a Local Green Space. This is where the tiered approach to our spatial strategy could come into effect as if we lose the Green Belt, then hopefully the garden village principles will be applied to ensure separation around the new settlement, and if that does not happen, then the buffers should/could come into effect, and if that does not happened, then the Strategic Green Gaps and Local Green Spaces should kick in. With each step further down the line we secure less and less green space and less and less separation, but we will have maximised our chances of securing at least some form and degree of separation.

What this shows is that the capacity to influence decisions does not depend solely on the strength of connections within the neighbourhood planning network (Gallent and Robinson, 2013), nor does it depend solely on the capacity of the local authority to listen, represent, negotiate and act between the state and the community (Parker, 2012; Wills, 2016; Parker and Salter, 2017), although both of these are important. In our case the capacity to influence decisions might be far more dependent upon the presentational strategy in both written and oral form and – most significantly – on the formal power invested in roles and individuals higher up the planning hierarchy. Indeed, in circumstances where there is either no active negotiation between the state and the community, or where there is negotiation but this is obstructive to a neighbourhood plan, involvement in processes such as the public examination of a local plan might prove particularly fruitful as the inspector has the power to direct the local authority to revisit issues of concern to the community as enshrined in the neighbourhood plan. It also shows that although the power to set principles remains with the LPA (Gallent and Robinson, 2013) this power is not immune to other influences, especially if those influences are themselves framed at the level of principle. In a classic example of the local jumping scales (Brownill, 2017), the neighbourhood here can jump to the level of influence invested at the national level through the examination process. In this instance, it was not the neighbourhood plan itself but the strategic nature of its contents and its presentation that seems to have carried weight. In essence, then, it is only by being what it is not intended to be (strategic) that a neighbourhood plan is likely to have any influence on an emerging local plan and even then, it requires national direction and authority for any potential influence to be realised.

Regrettably, irrespective of any creative potential we might have carved out with respect to the strategic capacity of neighbourhood planning, this example merely reinforces the lack of genuine power that is transferred to the people within neighbourhood planning as any influence we might ultimately have on the local plan will only have been secured by the intervention of the national level authority of the planning inspector.

Conclusion

This chapter highlights the need to think beyond the production of the plan as a finished document to consider how neighbourhood planning can be integrated with other planning processes at higher levels to secure greater weight for community views and greater likelihood that those views will be respected in the planning process. While I would agree that it is important to identify the policy space early in the process in order to manage expectations (Bradley and Brownill, 2017), it is also important to scope out the more-than-policy space beyond the neighbourhood plan itself to identify and maximise the potential to shape a local plan with which a neighbourhood plan will ultimately need to conform. In other words, we need to redefine the place of neighbourhood planning within the broader structure and process of plan making: neighbourhood planning does not need to come below, after and in subservience to the local plan if the development of a neighbourhood plan is strategically timed to facilitate community influence on emerging priorities and principles at the local plan level.

Such an approach does risk upsetting relations between the community and the LPA but as has been identified in the literature, false consensus and evading conflict are not necessarily beneficial (Parker et al, 2017; Vigar et al, 2017), and they certainly cannot be if that renders neighbourhood aspirations and concerns unaddressed. This after all is the whole point of neighbourhood planning and I suspect is the reason why most of us either became involved in the first place or stuck with it when the going got tough. The manoeuvrings between a neighbourhood planning team and an LPA – whether overtly political or more subtly practically oriented – are themselves revealing of the latent potential within a reconfigured place of neighbourhood planning as local authorities would surely not feel the need to employ such tactics (if indeed they do) if a neighbourhood plan was genuinely powerless.

Our participation at the public examination of LPP2 was very much a performance and the presentational construction of place both in the spatial strategy within the neighbourhood plan and in the articulation of community concerns at the 'in-principle' level are mutually supportive in seeking to optimise the chances of our aspirations for our neighbourhood being performed into reality through the local and/or neighbourhood, plan. It is possible, then, to perform neighbourhood planning differently and create new political opportunities from within the existing system (Parker and Street, 2015; Wills, 2016; Etherington and Jones, 2017) in the hope of influencing the principles laid out in a local plan by articulating the emerging priorities of a neighbourhood plan at the level of principle or strategy. By presenting residents' and communities' concerns for their area in a different – more strategic – way, new opportunities become available to widen the sphere of influence for a neighbourhood plan. However, this ultimately requires a neighbourhood plan to free itself from its own definition and position within the planning system and still relies upon other, higher, formalised authority to have any real impact. Consequently, while there is scope to be creative in the doing of neighbourhood planning in its current guise, I concur with others who argue for a much more radical reformulation of what a neighbourhood plan is, what it can do, where it fits within the planning system, and how it relates to the local plan (Parker et al, 2017; Parker and Salter, 2017; Wargent and Parker, 2018). There is a definite need to remake the place of neighbourhood planning within the planning system as much as a neighbourhood plan is optimistically deemed to remake the place to which it refers.

References

Bradley, Q (2015) The political identities of neighbourhood planning in England. *Space and Polity* 19 (2): 97–109

Bradley, Q (2018) Neighbourhood planning and the production of spatial knowledge. *The Town Planning Review* 89 (1): 23–42

Bradley, Q and Brownill, S (2017) Reflections of neighbourhood planning: towards a progressive localism. In S Brownill and Q Bradley (eds) *Localism and neighbourhood planning: power to the people?* Policy Press: Bristol, p251–267

Brownill, S (2017) Assembling neighbourhoods: topologies of power and the reshaping of planning. In S Brownill and Q Bradley (eds) *Localism and neighbourhood planning: power to the people?* Policy Press: Bristol, p145–161

Brownill, S and Bradley, Q (2017) Introduction. In S Brownill and Q Bradley (eds) *Localism and neighbourhood planning: power to the people?* Policy Press: Bristol, p1–15

Etherington, D and Jones, M (2017) Re-stating the post-political: depoliticization, social inequalities, and city-region growth. *Environment and Planning A: Economy and Space* 50 (1): 51–72

Gallent, N and Robinson, S (2013) *Neighbourhood planning: communities, networks and governance*. Policy Press: Bristol

Gunn, S, Brooks, E and Vigar, G (2014) The community's capacity to plan: the disproportionate requirements of the new English neighbourhood planning initiative. In S Davoudi and A Madanipour (eds) *Reconsidering localism*. Routledge: New York and London, chapter 8. (no page numbers) Accessed 6 Dec 2018

Kenis, A (2018) Post-politics contested: why multiple voices on climate change do not equal politicisation. *Environment and Planning C: Politics and Space* DOI:10.1177/0263774X18807209

MHCLG (2018) *Garden communities: prospectus* www.gov.uk/government/publications/garden-communities-prospectus Accessed 22 Jan 2018

Mouffe, C (2005) *On the political*. Routledge: Abingdon

Parker, G (2012) *Neighbourhood planning: precursors, lessons and prospects*. 40th Joint Planning Law Conference, Oxford www.quadrilect.com/Gavin%20Parker.pdf Accessed 15 Dec 2018

Parker, G (2017) The uneven geographies of neighbourhood planning in England. In S Brownill and Q Bradley (eds) *Localism and neighbourhood planning: power to the people?* Policy Press: Bristol, p75–91

Parker, G, Lynn, T and Wargent, M (2015) Sticking to the script? The co-production of neighbourhood planning in England. *The Town Planning Review* 86 (5): 519–536

Parker, G, Lynn, T and Wargent, M (2017) Contestation and conservatism in neighbourhood planning: reconciling agonism and collaboration? *Planning Theory and Practice* 18 (3): 446–465

Parker, G and Salter, K (2017) Taking stock of neighbourhood planning 2011–2016. *Planning Practice and Research* 32 (4): 478–490

Parker, G, Salter, K and Wargent, M (2019) *Neighbourhood planning in practice*. Lund Humphries: London

Parker, G and Street, E (2015) Planning at the neighbourhood scale: localism, dialogical politics, and the modulation of community action. *Environment and Planning C: Government and Policy* 33: 794–810

TCPA (2017) *Locating and consenting new garden cities: practical guide* www.tcpa.org.uk/guide-1-locating-and-consenting-new-garden-cities

TCPA (2018) *Understanding garden villages: an introductory guide* www.tcpa.org.uk/understanding-garden-villages

TDRC (2018) *Wootton and St Helen Without parishes character assessment* www.wshwnp.org.uk/wp-content/uploads/2018/08/WSHWNP-Character-Assessment-Introduction-full.pdf

Vigar, G, Gunn, S and Brookes, E (2017) Governing our neighbours: participation and conflict in neighbourhood planning. *The Town Planning Review* 88 (4): 423–442

VWHDC (2016) *Vale of white horse district council local plan 2031, part 1: strategic sites and policies* www.whitehorsedc.gov.uk/services-and-advice/planning-and-building/planning-policy/new-local-plan-2031-part-1-strategic-sites

Wargent, M and Parker, G (2018) Re-imagining neighbourhood governance: the future of neighbourhood planning in England. *The Town Planning Review* 89 (4): 379–402

Wills, J (2016) Emerging geographies of English localism: the case of neighbourhood planning. *Political Geography* 53: 43–53

WSHWNPSG (2018a) *Wootton and St Helen Without Neighbourhood Plan 2018–2031, consultation statement* www.whitehorsedc.gov.uk/sites/default/files/WSHWNP%20Consultation%20Statement_0.pdf

WSHWNPSG (2018b) *Wootton and St Helen Without Neighbourhood Plan 2018–2031* www.whitehorsedc.gov.uk/sites/default/files/WSHSNP%20Plan.pdf

WSHWNPSG (2018c) *Wootton and St Helen Without Neighbourhood Plan 2018–2031, screening opinion* www.whitehorsedc.gov.uk/sites/default/files/WSHWNP%20Screening%20Opinion.pdf

6 Reflections and projections

Introduction

Neighbourhood planning has been described as a means of enhancing place identity, meaning and attachment (Bradley, 2017a). On the one hand it has been argued that place attachment can provide a counter-narrative to NIMBYism as it posits the neighbourhood as a field of care (Bradley, 2017a), although conceptualising the neighbourhood as a field of care might equally be considered as motivation for NIMBYism. On the other hand, it has been suggested that strength of feeling can be so strong that rational deliberation can become difficult (Vigar et al, 2017), but this is by no means inevitable. Consequently, the same neighbourhood planning team might be equally perceived to have forged a legitimate and rational narrative around a field of care, or to have fallen into the irrational NIMBYist trap of reactionism. The possibility of a counter-narrative does not inevitably eradicate accusations of NIMBYism, and there remains no guarantee that local knowledge or place attachment will be incorporated into planning deliberations and outcomes.

Further, it is entirely possible that place attachment is not enhanced by neighbourhood planning but damaged by it. The presumed incommensurability of local place knowledge and formal spatial knowledge and the perceived threat to place attachment arising from major development proposals have been identified as potential frustrations to place attachment in neighbourhood planning (Bradley, 2017a, 2018; Lennon and Moore, 2018). However, other contributing factors include the unassailable prioritisation of the economic over the environmental and the social in the NPPF as argued in Chapter 3; the relegation of neighbourhood plans to consideration of inconsequential details after the substantive strategic work has been done, as explored in Chapter 4; and the formal power dynamics and softer strategies of process manipulation and policy modulation that can be brought to bear on neighbourhood planning teams by an LPA, as discussed in Chapter 5. These procedural factors, too, could significantly impact upon place identity and place attachment.

In this final substantive chapter, I work through three related critical concerns. I begin by exploring how we might conceptualise these negative experiences of place identity and attachment to critique the impact of neighbourhood planning, revisiting the potential for greater equivalence between community and planner experiences of the spaces and places of neighbourhood planning. Subsequently, I examine neighbourhood planning in terms of its legitimacy and prospects, developing critical and community perspectives on the legitimacy of neighbourhood planning, and propose modifications to the place, role and process of neighbourhood planning within the broader planning system to enhance its legitimacy, efficacy and efficiency.

Theorising place

In Chapter 2, I outlined a suggestion that the impacts of neighbourhood planning might be unsettling for those developing such plans as they continue to lead their phenomenologically immersed daily lives in their neighbourhood, the future of which has been rendered radically unstable. In this section, I investigate this suggestion more thoroughly in relation to the analytical conclusions from each chapter, working towards an innovative conceptualisation of the spaces and places of neighbourhood planning which encapsulates this instability.

In Chapter 3, I examined the different knowledges, languages, evaluation criteria and perspectives employed by different parties in neighbourhood planning. For example, LPP2 proposed to delete an area of land from the Green Belt but the neighbourhood plan articulated the community's desire to maintain the Green Belt, while LPP2's priority for sustainable development was considered to be the new settlement whereas we emphasised the importance of sustainability for existing settlements. These alternatives can come across in something of an 'all or nothing' fashion in the respective documents so that even though the site might be removed in part rather than in whole from the Green Belt and sustainable development might be delivered in something of a composite form, they are often perceived in oppositional, absolute terms. Until the planning process is complete, residents – especially those at the business end of neighbourhood planning – are stuck in limbo between two seemingly mutually exclusive possibilities. The Green Belt stays or it goes, the community wins or it loses. This can be both highly unsettling in terms of place identity and affectively challenging in terms of place attachment, as the neighbourhood is no longer the stable, familiar place that it previously was. As our expectations in terms of the likely outcomes of the neighbourhood planning process for our locality ebb and flow, so too can our place identity and attachment.

Similarly, in Chapter 4 – in relation to scalar, spatial and temporal approaches to strategy – oppositional binaries hove into view. The conventional view is that strategy is done at one level of the planning system but not the other; spatial strategies can be generic or specific; and either the local plan or the neighbourhood plan will be completed first. While our neighbourhood plan seeks to carve a strategic role for itself, we have no idea whether this will be upheld or quashed at examination, with implications for whether we can include both generic and site-specific policies. In this case, while we might try to do both elements in a binary or hybrid form, we might yet be relegated to doing only one. Again, the outcomes are uncertain and the experience unsettling as our expectations, confidence and place relations ebb and flow, fluctuating between oppositional possibilities.

In Chapter 5, I discussed the crucial relationship between neighbourhood planning teams and LPAs, exploring potential for the use of both formal political hierarchies and softer forms of power, and the importance of presentational, audience and ownership issues. As in the previous two chapters, place identity and attachment were subjected to contradictory influences. One day we were acting autonomously in drafting our plan but the next I considered that formal political hierarchies were being re-imposed to pressure our parish council 'bosses' to put us 'back in our place'. Similarly, the lack of clarity as to the most important audience for and ownership of a neighbourhood plan caused confusion for residents and frustration for the steering group. Once again, how we related to the place of our neighbourhood slid and slipped between radically different and sometimes directly contradictory subjectivities: autonomous, capable agents acting legitimately contrasted with supposedly inept or clueless troublemakers.

Place attachment, then, is not always or necessarily a constant source of positive affect but is both variable and vulnerable. It can be damaged as much as it can be nurtured, and it can be affected for better or for worse by influences on both the place to which a person is attached and the person who is attached to the place. How, though, might we conceptualise this affective dimension of neighbourhood planning, this unsettling sense of being stuck in limbo, suspended between uncertain outcomes?

Geography's interest in and abiding battle with binaries gives us an entry point into this issue as it has for some time now employed the idea of the hybrid to accommodate socio-natural combinations (see, for example, Whatmore, 2002; Lulka, 2009) but has also more recently started to engage in a sustained fashion with liminality (see, for example, McConnell, 2017; McConnell and Dittmer, 2018; Leyshon, 2018). The hybrid is characterised by the combination or merger of two entities in one, in a case of being both things simultaneously (Severi, 2015). The liminal is characterised by a stage in a ritual progression from one state to another during which the entity doing the progressing has left the first state but has not yet joined the second state, in a case of being neither one thing nor the other yet both at the same time (van Gennep, 1960; Turner, 1967, 1969, 1982). Unfortunately, neither of these concepts fits the preceding examples very well as they fail to capture the instability or fluctuating nature of the experiences described. In many cases it is simply not possible to be both of two possibilities simultaneously (e.g. a neighbourhood plan cannot be finished both before and after a local plan) or to be neither of two possibilities simultaneously (e.g. an area of land is either in or out of the Green Belt). Even in instances where it might be possible to be or do two things concurrently (e.g. strategic and detailed), it also might not. The whole point is that we do not know, and it is the not knowing that generates the radical fluctuations in both personal and place identity and place attachment.

Seemingly, then, we need a new way of thinking about situations in which something faces the prospect of being one thing or another, that cannot be both and cannot be neither, and in which the outcomes are unknown, generating a sense of flickering or fluctuating between alternatives. Instability is a good place to start as this speaks to the uncertainty and potential for changeability, but instability does not necessarily entail flickering or fluctuating as an unstable item might simply topple one way and come to rest. As a qualifier, my term of choice is chimerical, drawing on the idea of the chimera as something that – while also entailing two (or more) parts – is not fully any one of its component parts (MacKellar and Jones, 2012). Whereas with the hybrid two things come together so that each is more than itself, with the chimera two things come together such that neither is all of itself. Developing a line of thinking that emerged in relation to practice-based research, my sense of the chimera in this context is that it cannot be identified as one of its component parts (e.g. a strategic document) before the other component imposes itself (e.g. the expectation not to be strategic), generating an essential instability between different possibilities (Banfield, 2017, 2018). Applied to neighbourhood planning, as soon as we start performing according to optimistic possibilities something will happen to prompt us to flip into pessimism, suddenly feeling delegitimised and displaced.

Place attachment, though, is just one aspect of place. Previous chapters have also attended to the place of neighbourhood planning within the broader planning system, the space of the neighbourhood plan as a document and the place of the neighbourhood itself. A detailed analysis of how these spaces and places relate to this tripartite conceptualisation is presented in Table 6.1.

With respect to the place of neighbourhood planning within the broader planning system, this can be characterised as either liminal or chimerical. In some instances, the neighbourhood plan is trying to be or do two things at once (e.g. strategic and detailed, expert and lay knowledges) while simultaneously not being either in a pure form, thereby demonstrating liminality between the expectations of a professional strategic local plan and a community-based detailed neighbourhood plan as it seeks to span two different positions or roles in the planning hierarchy. In other cases, the neighbourhood plan can very definitely be or do only one thing or the other (e.g. be completed first or last, influencing the Green Belt or not doing so), thereby demonstrating chimericality.

In terms of the neighbourhood plan as a document, this is best characterised as hybrid in that – given the uncertainty of outcomes – the prudent approach is to cater for both extreme possibilities and thereby anything in between. The neighbourhood plan therefore incorporates both expert and lay language, pushes for community aspirations but provides for defensive measures if unsuccessful, and employs both strategic and detailed policies. In this case, it seeks to be or do both and is therefore hybrid, although some of these efforts might be thwarted at examination.

The place of the neighbourhood itself is less easy to characterise, as although the positions of the local plan and neighbourhood plan are often framed in oppositional terms, the implications of neighbourhood planning 'on the ground' are more likely to be a mish-mash of some outcomes that the community wanted and some outcomes that the LPA and developers wanted. That said, though, there are some aspects which remain chimerical, as there can only be one plan that is completed first and decisions taken at examination might dictate that we can do

Table 6.1 Theorising the spatial experiences of neighbourhood planning

Chapter	*Sub-section*	*Binaries*	*Place attachment*	*Place of neighbourhood planning*	*Neighbourhood plan document*	*Place of the neighbourhood*
Thematic	Environment	Expert vs lay knowledge	Chimerical Not knowing which will 'win': fluctuating options.	Liminal Neither fully one nor the other, but also both concurrently.	Hybrid NP caters for both extremes: abstract net gain and situated particularities.	- The outcome could be either extreme or anything in between.
	Green Belt	In vs out of the Green Belt	Chimerical Not knowing which will 'win': fluctuating options.	Chimerical NP either will or will not influence treatment of the GB.	Hybrid NP caters for both: the tiered strategy covers both extremes.	Chimerical/hybrid The outcome could be a totalising in/out decision, or a bit of both.
	Sustainable Development	New development vs old settlements	Chimerical Not knowing which will 'win': fluctuating options.	Chimerical The NP either will or will not influence the LP's approach.	Hybrid The NP caters for both possibilities.	- The outcome could be either extreme or anything in between.
Strategic	Scalar	Strategy vs detail	Chimerical Not knowing which will 'win': fluctuating options.	Liminal Neither fully one nor the other, but also both concurrently.	Hybrid The NP delivers both strategy and detail, or at least it aims to do so.	Chimerical/hybrid The outcome could be that we can only influence detail, or a combination of strategy and detail.
	Spatial	Generic vs site-specific	Hybrid Place attachment is both generic and site-specific.	Chimerical/hybrid The relationship between generic and strategic is unclear, but cannot be neither.	Hybrid The NP delivers both generic principles and site-specific policies.	Hybrid No direct challenge to the NP doing both, providing the generic not associated with the strategic.
	Temporal	NP before vs after LP	Chimerical Uncertain which will get there first, but only one can.	Chimerical Uncertain which document will get there first, but only one can.	Hybrid The NP seeks to cover both scenarios.	Chimerical Outcomes highly dependent upon which document gets there first.

(Continued)

Table 6.1 (Continued)

Chapter	*Sub-section*	*Binaries*	*Place attachment*	*Place of neighbourhood planning*	*Neighbourhood plan document*	*Place of the neighbourhood*
Performative	Political	Power over vs power to	Chimerical Fluctuation between legitimacy and illegitimacy.	Chimerical Perceived LPA tactics hint at possible shift in NP-LP relation; unlikely given power dynamic.	Hybrid The NP tries to balance community needs and LPA demands so process continues.	Chimerical Outcomes highly dependent upon which document gets there first.
	Practical	Helping vs obstructing	Chimerical Fluctuation between NP agency and lack of agency.	Chimerical Fluctuation between needing help and fearing domination.	Hybrid NP tries to balance responding to advice and avoiding possible interference.	Chimerical/hybrid Likely a mix of genuine and motivated 'help'.
	Presentational	Clarity of ownership vs lack of clarity	Chimerical Fluctuation between self-directed and LPA enrolment.	Chimerical/liminal Fluctuation between NP as a community or LPA production.	Hybrid NP tries to balance system needs and community needs.	Chimerical/hybrid NP could meet only community or formal needs, but likely a mix.

only one thing (e.g. detail) and not another (e.g. strategy), which will determine the future of the neighbourhood in unilateral ways.

Reading across the three analytical chapters, then, we can find instances that are best described as hybrid (being two things simultaneously), others that fit description as liminal (being both and yet neither thing simultaneously), and yet others that can be characterised as chimerically unstable (reliably being neither one thing nor the other but fluctuating between the options). The impacts of neighbourhood planning on place identity and place attachment are best considered chimerical; the place of neighbourhood planning is either liminal or chimerical; the space of the neighbourhood plan as a document is hybrid, and the place of the neighbourhood can be variously described as any or all of those terms. The ultimate outcomes for place identity and place attachment will depend upon the specific constellation of possibilities that comes to pass and might be predominantly positive or predominantly negative. Along the way, though, is a host of opportunities for highly destabilising spatial experiences, and the hybrid, the liminal and the chimerically unstable provide three related terms with which to conceptualise the varied spatialities of neighbourhood planning.

Developing this spatial conceptualisation further, it becomes clear that the multiple binaries characterising our experience of neighbourhood planning generate multiple hybrids, liminalities and chimerical instabilities, which can overlap, intersect and nest with one another due to the complex interconnectivities between the neighbourhood, the neighbourhood plan and planning more broadly. For example, the emphasis placed on expert compared to local knowledge at examination might influence the interpretation of sustainable development, what our plan can therefore say and how development unfolds in our neighbourhood. While this highlights both the complexity and potential scale of influences on place attachment, such conceptualisations can be utilised independently and in combination, affording both breadth in terms of contextual relevance and specificity in terms of analytical application.

One final analytical point warrants discussion here: the equivalence that can be drawn between chimerical instability experienced by residents in relation to place attachment, and the chimerical instability generated in relation to the place of neighbourhood planning within the broader planning system. Picking up on synergies rather than distinctions between lived place and abstract space, each can be conceptualised and experienced in a similarly unsettling manner, as outlined in Chapter 2. In our case, both were incidental effects of the neighbourhood planning process. Our attempts to be strategic and to get ahead of the local plan certainly generated additional uncertainties for us in destabilising place identity and attachment, but it also generated uncertainties for every actor in the planning process because it raised the possibility of a neighbourhood plan doing things that it normally would not do. The prospects were raised of a neighbourhood plan being finalised ahead of a local plan, having something to say about strategy, being able to influence outcomes for the Green Belt and sustainable development, all of which brought the potential to cause difficulties for the LPA in the context of dwelling practically among their plans just as we encountered difficulties in dwelling practically in our neighbourhood. The critical point here is that whereas for us the chimerical instability might have been an incidental side-effect (detrimental for our place attachment but potentially productive for community outcomes), the possibility exists to employ chimerical instability as a tactic in the armoury of neighbourhood planning teams. By injecting and sustaining chimerical instability in the planning process, alternative possibilities and outcomes can be held open and kept in view. The tactical potential of chimerical instability thereby lies in helping to prevent possibilities that are beneficial for the community from being shut down prematurely. It might frustrate community-LPA relations, and it might make life very uncomfortable, but it does so for both the community and potentially the LPA, and it might increase the chances of community objectives being delivered, however minimally.

Legitimacy

The specific purpose of neighbourhood planning to reduce popular resistance to development and increase housebuilding is widely acknowledged and has previously been critiqued (Gallent and Robinson, 2013; Haughton et al, 2013; Cowell, 2015; Bradley, 2017a, 2017b; Bradley and Brownill, 2017; Brownill, 2017; Colomb, 2017; Parker, 2017; Vigar et al, 2017; Wargent and Parker, 2018; Wills, 2016). Less critical attention seems to have been directed, though, towards the broader purpose of planning, perhaps symptomatic of the proportion of academic literature on neighbourhood planning being produced by professional planners who have been socialised into and therefore perhaps tend not to question the underpinnings of their own profession (Stephenson, 2010). This issue, though, is an important starting point for discussion of legitimacy.

It seems unproblematic to suggest that the overriding concern of the planning system is the effective direction and shaping of development (Parker, 2012) but difficulties arise when phrases such as effective direction and shaping are accepted unquestioningly. Similar difficulties arise when questions are posed as to whether policy vehicles such as neighbourhood plans will improve planning outcomes or shape place qualities to promote better development trajectories (Parker, 2012; Healey, 2015) without exploring what constitutes place quality, improved planning outcomes or development trajectories, and who gets to define them. Assertions that what is needed is to develop a better popular understanding of the need for and benefits of development or to improve the intelligence of the polity (Parker, 2012; Healey, 2015) are also problematic, as they delegitimise differences of opinion irrespective of their validity. Such assertions seem to me to suggest that if clever planners explain development to stupid people then people would be less stupid and all development would correctly be deemed advantageous, thereby eradicating resistance. However, such views overlook the fact that not all development *is* needed for the public good, and that what constitutes the public good is just as contentious as the evidence base on which it is established, especially in circumstances where redistributive benefits or any sense of social justice appear to have been jettisoned. There is far more at stake here than simply educating supposedly stupid people, as the broader planning project itself has critical questions to consider in relation to its own legitimacy. This is especially so given the ecomomically oriented operationalisation of sustainable development in the NPPF and the known difficulties in ensuring the delivery of affordable housing due to the supremacy of the interests of the development industry despite the efforts of neighbourhood planning teams to increase provision (Parker et al, 2017; Wargent and Parker, 2018). Unfortunately, such critical consideration would probably be deemed too strategic for lowly volunteer neighbourhood planners to comprehend, who are presumably considered intellectually closer to stupid people than clever planners, but those involved in neighbourhood planning have a specific and informed perspective on such critical questions and could legitimately contribute to their deliberation, if only they were allowed and invited to do so.

Irrespective of these broader issues, though, how might we evaluate the legitimacy of neighbourhood planning itself? A starting point for this discussion is an embryonic framework of legitimacy – the transformation of power into authority – in which legitimacy can be either descriptive (if deemed to be appropriate, e.g. based upon legality or tradition) or normative (if deemed to be morally justified), although these two often overlap (Cowie and Davoudi, 2015). Within this framework, three grounds for considering authority to be legitimate are identified: 1) a person or group gave initial *consent* to its establishment, 2) a person or group consider the *procedure* to be legitimate as distinct from the outcome, and 3) a person or group base legitimacy judgements on whether the *substantive* outcomes are beneficial for them (Cowie and Davoudi, 2015). Procedural legitimacy is then subdivided further into representative and participatory dimensions, and the representative dimension is categorised into four different types: formalistic, in which one person speaks for another; symbolic, in which one person stands for another; descriptive, in which one person resembles another, and substantive, where one person acts for another.

Within this framework, neighbourhood planning did not fare well in terms of legitimacy (Cowie and Davoudi, 2015) because, among other things:

1. The Coalition government that introduced it did not have overall control and therefore consent is questionable.
2. The legitimacy of speaking for the community is undermined by the self-selecting nature of volunteers.
3. Symbolic legitimacy is questioned due to the low turnout at referendum.
4. Descriptive legitimacy is problematic due to the lack of representativeness among participants.
5. The substantive legitimacy of neighbourhood planning is hard to evaluate as the winners and losers, gains and losses, are hard to quantify.

While this framework and critical reading of neighbourhood planning are helpful in specifying the complexity of legitimacy evaluations and some of the shortcomings of neighbourhood planning, it also only provides part of the picture. Consent, for example, focuses on the national level rather than the local level, while only the representative aspect of procedural legitimacy is considered in detail, leaving the participatory aspects unexplored, and substantive legitimacy is equally undeveloped. Here, I develop this framework further based on my own experience of neighbourhood planning, contributing an alternative, 'community insider's' perspective, as specified in Figure 6.1. In this revised framework, I have retained the either/or combination of descriptive and normative dimensions of legitimacy in recognition that they might work in isolation or in combination with each other. I have also carried over the four-way categorisation of representative legitimacy to participatory legitimacy, to provide analytical consistency under the banner of procedural legitimacy. However, the new and expanded contributions to the original framework provide both additional details and countervailing perspectives in the hope of extending and deepening our engagement with issues of legitimacy in neighbourhood planning.

The modifications and additions of greatest significance are:

> *The distinction drawn between consent at the national level and the local level.* Concerns over the legitimacy of initiatives introduced in conditions

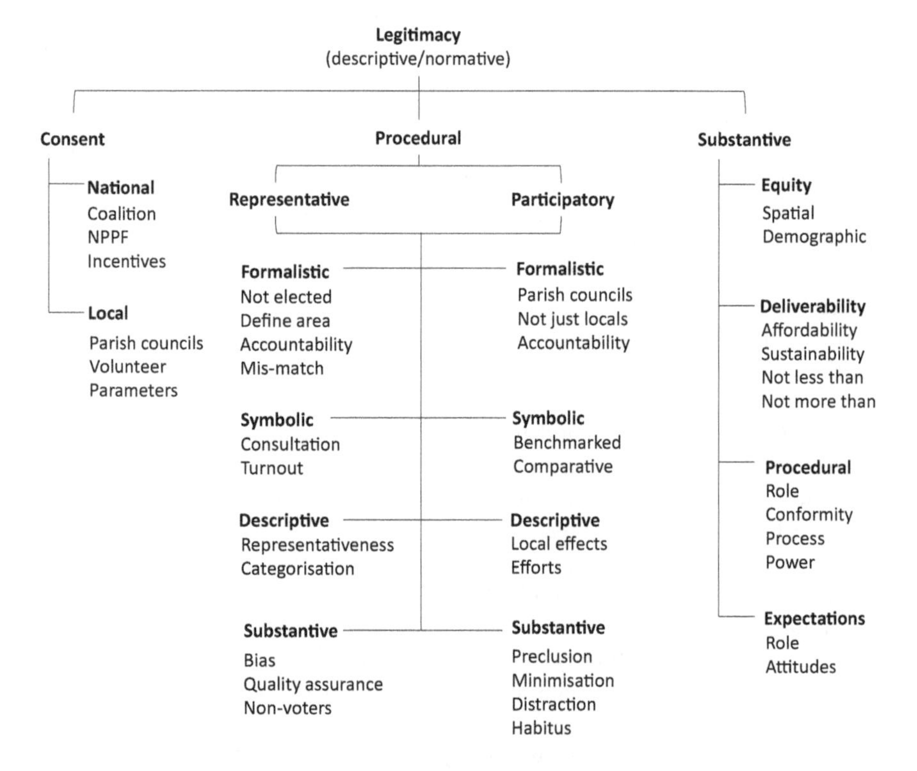

Figure 6.1 Legitimacy framework for neighbourhood planning

Source: Developed from Cowie and Davoudi, 2015

of no overall control are certainly valid, but there are also issues of consent at the local level, as a parish council or an embryonic community group is required to establish a steering group or neighbourhood forum, while consent is also needed on the part of the volunteers. In addition, the sustainable development stipulations of the NPPF raise concerns over legitimacy, as do the financial incentives made available to encourage participation and stifle opposition to development, as both factors mitigate against full and free consent. Further, the lack of detailed information about the nature, scale and burdensomeness of the process undermines volunteers' ability to make informed decisions as to what it is precisely to which they are consenting.

Additional detail for procedural (representative) legitimacy. There are undeniably issues to unpick concerning the self-selected nature of volunteers and the lack of representativeness that ensues, the low turnout for referendum and so on. However other issues also arise, such as the mismatch in formal representative democracy whereby residents of one location might belong to one parish but be covered by the neighbourhood plan of another or might not be covered by their own parish's neighbourhood plan. Equally, while the biases and power inequalities in the system favour the already powerful, the role of quality assurance measures, such as Basic Conditions Statements and SEA/HRA requirements, in helping to shape substantive outcomes also warrants consideration as their demands feed into those inequalities.

Additional and qualifying detail for procedural (participatory) legitimacy. While volunteer self-selection, Designated Area determination and the risk of unelected parish councillors can be problematic, I think we need to be careful not to overstate the case in these regards. For example, the determination of the Designated Area is subject to greater accountability than is sometimes assumed, as although the team behind a neighbourhood plan can propose a Designated Area, this proposal is open to challenge to people both within and beyond the area proposed. Similarly, while consultation response rates and referendum turnouts might be disappointing, we should avoid conflating broader issues of declining democratic participation with the specific shortcomings of neighbourhood planning, especially as turnouts for neighbourhood referenda are higher than for other elections (Bradley, 2017b; Bradley and Brownill, 2017). However, we also need to be aware of all those performative factors identified in Chapter 5 that mitigate against neighbourhood planning teams from acting for their community or from being seen to act for their community. By remaining alert both to situational particularities such as highly transient populations (who might be less inclined to participate) or opacity regarding neighbourhood plan ownership (which might reduce turnout), and to broader systemic issues such as declining civic participation, we are better equipped to attend to and address any issues that are uniquely attributable to neighbourhood planning.

Additional detail for substantive legitimacy. The difficulties in identifying and quantifying the gains and losses in relation to the actors involved has

> already been acknowledged (Cowie and Davoudi, 2015), but other difficulties arise here too. In our case, spatial inequalities complicate this calculation of gains and losses, as despite the neighbourhood planning team making the same arguments about the need for separation from the proposed development in relation to two neighbouring settlements, separation was provided for one settlement but not the other. Beyond these spatial inconsistencies, deliverability of substantive community gains is inevitably hampered by factors such as the dominant large-scale, speculative model of development, the 'economy first' focus of sustainable development in the NPPF, and the formal constraints that prevent the promotion of less development or more environmental protection measures than is specified by higher plans and policies. Structural issues include the constrained role of neighbourhood planning, the power imbalances, and the need for conformity with higher level plans. The concerns surrounding expectations spring from the discussion of performative understandings of place in the previous chapter, as volunteers choose to participate in initiatives such as neighbourhood planning because they expect to be able to do certain things, and when it turns out that these cannot in fact be done, the perceived legitimacy of the process evaporates.

This significantly reworked and expanded framework, then, acknowledges the interdependency between procedural and substantive legitimacy by presenting each as a sub-category of the other, and is more comprehensive in its coverage of the different types of legitimacy and more balanced in its perspective in covering national and local, formal and volunteer perspectives than its original presentation (Cowie and Davoudi, 2015). It attends to structural constraints, both hard and soft expressions of power, and interpersonal and performative aspects of neighbourhood planning. It also acts as a cautionary reminder that whatever shortcomings we might identify in neighbourhood planning, especially in terms of representativeness, self-determination, community engagement and turnout, neighbourhood planning does not take place in a vacuum but is part and parcel of a broader planning project, hierarchy and system. While we need to uncover and address the shortcomings inherent to neighbourhood planning, we must not lose sight of the broader context within which it takes place.

Unsurprisingly, my evaluation of the legitimacy of neighbourhood planning is fairly damning. The question of consent is murky at best as: free and informed consent are elusive; the proclamations and promises in relation to local autonomy, knowledge and place attachment appear to be baseless because the system excludes the very local knowledge and sentiment that it claims to seek; the NPPF is driven by economic growth rather than genuinely sustainable development, and the habitus of both planning as a profession and local government as an institution appear resistant to the stated aims of localism. Procedurally, neighbourhood planning should fare well as it is prescribed in some detail in law, has a variety of checks and balances in place and on paper at least, looks as though it should be able to incorporate local knowledge and place attachment. However, in terms of representative legitimacy the local is crowded out not only by the host of other actors involved in the process, all of whom hold more power than the community, but also by the formal hierarchy of power that can be brought to bear on the neighbourhood planning team by an LPA, the constraints imposed by the NPPF that close down local concerns, and the embedding of neighbourhood planning in a broader process that minimises the role of neighbourhood planning. Participatory legitimacy is also undermined as although communities can and do produce their plans and can and do feed into broader planning processes such as the examination of LPP2, these same power dynamics, institutionalised vested interests and structural hierarchies quash any meaningful opportunity for a neighbourhood to play a substantive role. Finally, in terms of substantive legitimacy, even if we were successful in securing a strategic role for our neighbourhood plan, in retaining the Green Belt, or in guaranteeing settlement separation, it is highly questionable whether this would be attributable to the neighbourhood plan. Even if we did attain a hypothetical utopia in which every community objective is delivered, whether through the neighbourhood plan itself or through any influence that we might turn out to secure on the emerging LPP2, this would be down to the relevant inspector/s for their respective examinations and the power that they had wielded on our behalf. In short, then, the legitimacy of neighbourhood planning in its current form, based on my own experience, is shot to pieces.

Prospects

Over the past few years, literature on neighbourhood planning has started to call for a more progressive form of localism. I engage with just two such calls here to set the scene for the discussion that follows. One seeks a stronger orientation to the future, concern for sustainability for the many rather than the few, a focus on interconnectivities across space and time, more open and transparent processes, and increasing the intelligence of the polity (Healey, 2015). The other promotes more equitable plan making in terms of geographic spread, greater co-production between the community and the LPA, greater social inclusion, improved quality and value added, reconciliation of local and strategic concerns, and enhanced community control (Wargent and Parker, 2018).

There are aspects of each of these that are attractive to me. The call for a stronger orientation to the future and sustainability for the many rather than the few (Healey, 2015), for example, chime nicely with my own sense both that neighbourhood planning should be invited into the realm of strategy and that sustainability must not be delivered at new developments at the expense of existing settlements. Such measures would help to engender a longer-term perspective at the level of the community than is permitted by the current system, alongside a broader spatial perspective on the part of planners and developers than is currently promoted by the NPPF's emphasis on sustainable *development* rather than sustainable settlements and communities. Similarly, calls for greater co-production between the community and the LPA, reconciliation of strategic and local concerns, and enhanced community control (Wargent and Parker, 2018), could all be progressed by reconfiguring the planning system to enable and encourage more strategic participation on the part of the community and greater sensitivity to locality on the part of the LPA.

Equally, though, alarm bells ring in relation to aspects of each of these accounts. While improving quality and added value (Wargent and Parker, 2018) seem self-evidently laudable aims, the same political questions arise as we saw with improving the development trajectory through planning: what forms does quality take, what value is added and for whom? However, increasing the intelligence of the polity (Healey, 2015) is the aspect that is potentially most troubling, as it depends rather on what is considered to constitute the polity. If the polity is confined to local people, residents and community volunteers, then we end up back in the situation where the people are considered stupid and the planners are their saviour, a perspective that is hardly conducive to collaborative working, let alone the incorporation of local knowledge into planning. If, on the other hand, the polity includes professional planners, site promoters, developers, interest groups and so on, then we find ourselves in a more democratic situation in which the perspectives and ambitions of all parties can be recognised and the capacity of each party to listen to, accept and accommodate the needs of the others can be enhanced. In this latter case, we might start to move towards more consensual and equitable outcomes that share the benefits of development among all parties.

Consequently, these two areas of concern – the promise of integrating the strategic and the local, and the need for more consensual, equitable outcomes – form the main focal points for this section, supplemented by a final consideration of matters of process that would improve the efficiency and efficacy of neighbourhood planning even in the absence of more radical changes.

Integrating the strategic and the local

If initiatives such as neighbourhood planning are to work, we need to develop public trust in statutory agencies and statutory faith in the quality of work undertaken by members of the public (Owen et al, 2007). However, this will only be possible if the system within which they operate facilitates it, and – as I see it – both the NPPF and neighbourhood planning fail spectacularly in this regard, by denying, excluding, deleting or diluting community voices and concerns, and by constraining community participation to the most minimal and inconsequential elements of planning, resulting in the gains accruing to developers and the costs being born by communities because communities are excluded from strategic matters that fundamentally determine the distribution of gains and losses.

The extant literature is replete with calls for a reconfiguration of the relationship between local and neighbourhood planning, for new ways to integrate the strategic and the local, and to reconsider the role and place of neighbourhood plans within the broader planning framework (see, for example, Parker, 2012; Gallent, 2014; Parker and Salter, 2017; Wargent and Parker, 2018). However, to ask how we might integrate the top-down and bottom-up is perhaps to ask the wrong question, as it presupposes the existence of two distinct bodies of information that need to be brought together. Instead, we need to rethink how the different levels of planning can be operationalised most effectively and ask how the strategic (top-down) can become more sensitive to the local and how the local (bottom-up) can become more strategic. After all, the strategic touches down somewhere and the local relates to the strategic in some way. The key then becomes how to enable and encourage each to inform the other before either of them becomes fully formulated independently.

My suggestion is that we move away from presupposed levels and roles of different plans to focus more on stages in the strategic process, with one set of discussions taking place at an abstract, unquantified level, and a second stage negotiating the specifics of how the first discussion is put into practice. The first discussion could be based on the in-principle level of planning, whereby an LPA determines where within its district it thinks development ought to be targeted and specifies why but without detailing dwelling numbers and exact locations, and invites communities (however defined) to consider where within their own sub-area they think development ought to be targeted and why, which might be strategically oriented (e.g. towards supporting smaller settlements) or locationally specific (e.g. towards one town or site) or some combination of the two. The LPA could then calculate its quantum of development across its district to reflect the preferences of communities and go back to those communities with a more locally sensitive outline development plan. This would allow the LPA to identify any shortfall or mismatches between development need and provision, prompting more tailored discussions at the community level, and would enable communities to confirm that their allocation preferences had been accommodated at an early stage. The specification as to the reasoning behind the stated allocation preferences would pave the way for recognition of different agendas and understandings, thereby supporting collaborative planning both within and between governance levels. Areas of disagreement could be subject to supplementary procedures to support consensual and equitable outcomes (for thoughts on which, see the next section), while the second discussion could also clarify and specify which aspects will be planned by the LPA and which will be planned by the community, generating a negotiated relationship and a process geared towards collaborative rather than adversarial planning. Such an approach could provide greater flexibility, enhanced sensitivity to locality, and greater community involvement in determining strategic issues, while both ensuring that development happens and capturing and developing community capacity for strategic decision-making.

How this could work in practice within the current planning system is a vexed question, not only due to the almost constant flux within planning as successive governments introduce their own changes, but also because a more plural approach such as this is necessarily more complex than a uniform set of procedures (Parker et al, 2017; Waite and Bristow, 2018; Wargent and Parker, 2018). Not being a planner myself, I cannot contribute much to the technicalities of how such a process might be orchestrated, but I am firmly of the view that the current system is not working well, and something needs to give. I am also firmly of the view that where there is a will there is a way, so although there would no doubt be difficulties encountered in such a radical change to how the lower tiers of planning work, this is no reason not to explore the options or to seek to resolve any difficulties that might arise (e.g. ensuring delivery on early commitments, providing a default process for localities that choose not to participate, or establishing arbitration measures). If as a nation, institution or profession, we have the intellectual capability to create a system that works so impressively well for one set of vested interests (the economic), we must surely be capable of creating a system that works impressively well in balancing diverse interests. Reconfiguring the relationship between the strategic and the local would be a good place to start.

Enabling consensual and equitable outcomes

The inequitable nature of neighbourhood planning in terms of both its procedural opportunities and its substantive outcomes is a major concern. As we saw in Chapter 3, instead of seeking cumulative variegated gain that could accommodate multiple and diverse interests, planning works on the basis of net (capital) gain, which is too crude to deliver locally sensitive outcomes.

Here, the issue of the 'economy first' model of 'sustainable' development is compounded by a seemingly implicit assumption on the part of those advocating development that economic viability is one and the same thing as maximum profitability, which narrows the space of negotiation into non-existence as anything that even minimally impacts on maximum profits is deemed to be economically non-viable. Consequently, despite aspirations that localism might herald a move away from development proposals being simply imposed on the people to development decisions being informed by multi-sided, multi-scaled debates (Healey, 2015), it seems to me that neighbourhood planning generates months and years of debate *and then* development is imposed on the people. The fact that we had a seat at the table is taken as proof that we had a meaningful say in the debate, but nothing could be further from the truth as I see it: we said an awful lot but very little of what we said appears to have been heard. Contrary to the purpose of neighbourhood planning being to enable communities to ensure that development in their area is appropriate to their community (Parker, 2012), any and all development is deemed appropriate because any arguments or evidence that would establish inappropriateness are dismissed, denied or excluded through the many and varied forms of epistemic violence investigated in this volume (e.g. the stipulations of the NPPF, the prioritisation of expertise, the formal imposition of institutional power, and so on).

What we need, it seems, is to be honest with ourselves about the failings of the current system, rather than hiding behind technical and administrative processes that claim to resolve them but do not. We need to recognise and respect the antagonistic basis of the political (Mouffe, 2005; Etherington and Jones, 2017; Kenis, 2018), but find a way to frame debate that allows it to play out more equitably and collaboratively. In this way, perhaps we can ensure that no one actor receives all the gains or suffers all the losses, so that the costs and benefits of planning and development are genuinely shared.

This does not mean that economic factors are ignored, but it does mean not allowing maximum profitability to stand in proxy for economic viability. Equally, it does not mean allowing communities to oppose any and all development, but it does mean taking seriously community concerns and taking meaningful steps to accommodate them. There must be a way of avoiding extreme outcomes for both sides and to ensure that no one party gets all the benefits while another makes all the sacrifices. This is reminiscent of the earlier discussion of hybrid, liminal and chimerically unstable places, in which both the experience of the neighbourhood on the part of community volunteers and the experience of the planning system on the part of the LPA were considered chimerical: deeply uncertain and radically unstable in fluctuating between oppositional potential outcomes. Living in a chimerically unstable neighbourhood is not much fun, and I suspect that living in a chimerically unstable planning system is not much fun either, so if we can find a way to use the idea of chimerical instability as a means of carving out a space of negotiation that reduces the spectre of experiencing chimerical instability, so much the better.

As a step in this direction, we also saw earlier that whereas the space of the neighbourhood plan was characterised as hybrid as it sought to accommodate both extreme possibilities, the place of the neighbourhood itself was more commonly a combination of chimerical, liminal and hybrid, as the possible combination of outcomes on the ground was far more multiple and varied than the binaries around which the neighbourhood plan as a document was implicitly based. What we are left with, then, is a situation in which the possible outcomes on the ground are innumerable but the positionings, framings and strategies tend to be oriented around binaries. It should therefore be possible to identify what the primary binaries are and to establish parameters for discussion that guarantee that no one party will have all the gains or all the losses; that neither their worst nor best case scenario will be realised on any of the binaries. Diagrammatically, this might look a bit like Figure 6.2.

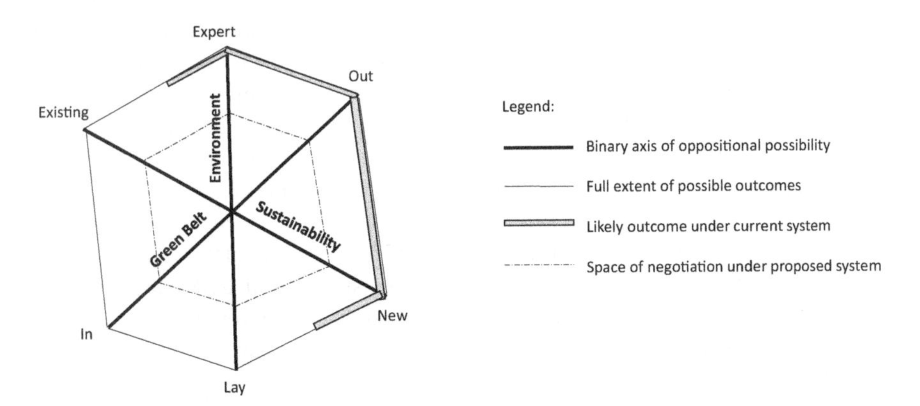

Figure 6.2 Chimerical instability delimiting a space of negotiation

Source: Author

The benefits of delimiting a space of negotiation among the multiple binary tensions are three-fold. First, it reduces the excessive dominance of any one vested interest so that the whole process is inherently more equitable as every party gives a bit and gains a bit. Secondly, it reduces uncertainty, as the most extreme possibilities are jettisoned before negotiation begins, enhancing the sense of security for all parties. Thirdly, because it is less extreme, it is also less affectively unsettling for all parties, which in turn is more conducive to continued collaborative planning, helping to bolster the reciprocal trust that is needed if it is to be effective (Owen et al, 2007).

Clearly, collaborative planning will not materialise if the different parties are left to self-regulate because the parties are not equally endowed with power and the whole system is stacked in favour of development and against communities. Developing a formalised mechanism for delimiting the parameters of the debate – the space of negotiation – is one way in which the different understandings, knowledges and agendas can be recognised while simultaneously working towards an approach to development in which the gains and losses are shared and mutually agreed, moving us that bit closer to a collaborative planning system.

Matters of process

In addition to wholesale reformulations of the planning system, a host of more minor adaptations to the procedural and technical specifics of neighbourhood planning could help to make it more efficient and effective. One suggestion raised in the literature is that neighbourhood planning should not have statutory status as it risks binding neighbourhood planning too tightly to state agendas (Parker, 2012), but it has also been acknowledged that statutory status is a significant motivating factor, enhancing both collective identification and effectiveness (Bradley, 2015; Parker et al, 2017). From my perspective, the statutory status of neighbourhood plans is the main factor in their favour because it suggests that however minimal your achievements might be, they are at least – in principle anyway – sacrosanct. Unfortunately, the scope of neighbourhood planning is so constrained that statutory status in this context is next to meaningless, but I think this is a different issue to the binding of neighbourhood planning to state agendas. If neighbourhood plans become more strategic and if communities are engaged in the formulation of state agendas as applied to their locality as legitimate strategic contributors, then statutory status will become more meaningful and state agendas less problematic because they will be more locally sensitive.

Irrespective of statutory status, there is a need for greater clarity and accountability regarding the respective roles and responsibilities of LPAs and communities, including the need for institutional oversight of neighbourhood planning activities (Parker et al, 2017). However, where responsibility for this oversight should lie is less clear, as the state, local authorities, developers and communities all have vested interests. There is also the need for institutional oversight to cut both ways, especially given the discussion in the previous chapter of the ways in which power can be brought to bear upon neighbourhood planning teams by LPAs, but the most efficient and effective way to reduce the need for formal oversight arrangements is to reduce the risk of grievances arising in the first place. Ensuring that the planning process operates on a more level playing field as suggested earlier would be a good first step in this direction.

It also seems to me that communities should have greater freedom to voice their support or otherwise for higher level plans and policies. If local communities are unable to say 'no' to what they do not want in their area but are also unable to say 'yes' to what they do want in their area, then what on earth can they do? In the interests of transparency, communities should be free to articulate their opposition in principle even if they are obliged by the regulations to support development in practice, but it seems especially nonsensical to deny local communities the right to voice their genuine support for favourable state policies. This would give more credence to the claims that neighbourhood plans are produced by a community for that community and would help to simplify and streamline the examination process.

Finally, I think there should be greater consistency in the statutory requirements for local and neighbourhood plans, as while both are subject to statutory consultation periods and public examination, only neighbourhood plans are required to go to referendum. While this increases both the length of the process and the procedural requirements for neighbourhood plans compared to local plans, it does allow communities to take a final view on any changes introduced through examination or by an LPA following examination, so could also play a valuable role in local planning as a step towards addressing the democratic deficit. Given the powerful voices of the development industry in planning, I can certainly see an argument for subjecting a local plan to referendum, and I would suggest that this is preferable to incorporating neighbourhood planning representatives into strategic bodies or partnerships (Wargent and Parker, 2018), which would do nothing to address the dominance of economic agendas or the crowding out of local and lay perspectives. Referendum requirements for local plans could perhaps work well within a reconfigured planning system to help equalise the planning processes and requirements, encourage greater accountability between the different planning bodies and enhance integration in the development of those plans.

Unless the system and process of neighbourhood planning, and the broader planning system within which it sits, is drastically reworked, I do not consider the prospects for neighbourhood planning to be good. While there are modifications that could be implemented to improve the process from within, these are inadequate without more radical overhauls of the NPPF to provide balance between different aspects of sustainable development, the relationship between local and neighbourhood planning to allow communities to act strategically, and the mode of engagement to facilitate a genuine move from adversarial to collaborative planning.

Conclusion

This chapter has developed the theoretical and practical implications of the analytical engagements with neighbourhood planning in the previous three chapters.

It has been established that place attachment is not merely a reserve of positive affect that can motivate and sustain participation and be enhanced through neighbourhood planning, but is fluid and vulnerable, amenable to being damaged by neighbourhood planning, and thereby risking the progressive erosion of the very place attachment upon which neighbourhood planning is predicated. The varied spaces and places of neighbourhood planning have been conceptualised in terms of liminality, hybridity and chimerical instability, which paved the way for the potential operationalisation of chimerical instability in two ways. On the one hand, chimerical instability as a tactic available to neighbourhood planning teams can be injected into the planning system to destabilise presupposed relations between local and neighbourhood plans, thereby keeping open many of the issues and debates that an LPA might want to close down. On the other hand, chimerical instability can be developed into a means of framing the space of negotiation in such a manner that extreme possible outcomes are avoided and more collaborative and equitable outcomes are assured as all parties give a bit and gain a bit.

Issues of legitimacy have also been addressed, expanding upon earlier work by providing a community perspective and incorporating into a framework of legitimacy the overriding critical points that emerged from the three analytical chapters, to establish that, in my opinion, the legitimacy of neighbourhood planning leaves much to be desired in relation to its consent, procedural and substantive forms. The prospects for neighbourhood planning formed the third main discussion in this chapter, with both minor and more radical suggestions proposed for developing neighbourhood planning into a process and practice that is more balanced, democratic, collaborative and legitimate than is currently the case. While issues of statutory status, process requirements and the duty to support are all in need of attention, the bigger issues of the NPPF, the relationship between local and neighbourhood planning and the way in which decision-making in planning is framed all need a more radical overhaul if neighbourhood planning is to be in any way effective or legitimate. Without such modifications, I fear that the prospects for neighbourhood planning, and indeed for neighbourhoods, are dire.

References

Banfield, J (2017) Researching through unfamiliar practices. *Cultural Geographies* 24 (2): 329–332

Banfield, J (2018) Challenge in artistic flow experiences: an interdisciplinary intervention. *Qualitative Research in Psychology* https://doi.org/10.1080/14780887.2018.1475535 Accessed 14 Jan 2019

Bradley, Q (2015) The political identities of neighbourhood planning in England. *Space and Polity* 19 (2): 97–109

Bradley, Q (2017a) Neighbourhood planning and the impact of place identity on housing development in England. *Planning Theory and Practice* 18 (2): 233–248

Bradley, Q (2017b) Neighbourhoods, communities and the local scale. In S Brownill and Q Bradley (eds) *Localism and neighbourhood planning: power to the people?* Policy Press: Bristol, p39–55

Bradley, Q (2018) Neighbourhood planning and the production of spatial knowledge. *The Town Planning Review* 89 (1): 23–42

Bradley, Q and Brownill, S (2017) Reflections of neighbourhood planning: towards a progressive localism. In S Brownill and Q Bradley (eds) *Localism and neighbourhood planning: power to the people?* Policy Press: Bristol, p251–267

Brownill, S (2017) Assembling neighbourhoods: topologies of power and the reshaping of planning. In S Brownill and Q Bradley (eds) *Localism and neighbourhood planning: power to the people?* Policy Press: Bristol, p145–161

Colomb, C (2017) Participation and conflict in the formation of neighbourhood areas and forums in 'super-diverse' cities. In S Brownill and Q Bradley (eds) *Localism and neighbourhood planning: power to the people?* Policy Press: Bristol, p127–144

Cowell, R (2015) Localism and the environment: effective rescaling for sustainability transition. In S Davoudi and A Madanipour (eds) *Reconsidering localism*. Routledge: New York and London, chapter 11. (no page numbers) Accessed 6 Dec 2018

Cowie, P and Davoudi, S (2015) Is small really beautiful? The legitimacy of neighbourhood planning. In S Davoudi and A Madanipour (eds) *Reconsidering localism*. Routledge: New York and London, chapter 9. (no page numbers) Accessed 6 Dec 2018

Etherington, D and Jones, M (2017) Re-stating the post-political: depoliticization, social inequalities, and city-region growth. *Environment and Planning A: Economy and Space* 50 (1): 51–72

Gallent, N (2014) Connecting to the citizenry? Support groups in community planning in England. In N Gallent and D Ciaffi (eds) *Community action and planning: contexts, drivers and outcomes*. Policy Press: Bristol, p301–322

Gallent, N and Robinson, S (2013) *Neighbourhood planning: communities, networks and governance*. Policy Press: Bristol

Haughton, G, Allmendinger, P and Oosterlynck, S (2013) Spaces of neoliberal experimentation: soft spaces, postpolitics, and neoliberal governmentality. *Environment and Planning A* 45: 217–234

Healey, P (2015) Editorial. *Planning Theory and Practice* 6 (1): 5–8

Kenis, A (2018) Post-politics contested: why multiple voices on climate change do not equal politicisation. *Environment and Planning C: Politics and Space* DOI:10.1177/0263774X18807209

Lennon, M and Moore, D (2018) Planning, 'politics' and the production of space: the formulation and application of a framework for examining the micropolitics of community place-making. *Journal of Environmental Policy and Planning* DOI:10.1080/1523908X.2018.1508336

Leyshon, C (2018) Finding the coast: environmental governance and the characterisation of land and sea. *Area* 50: 150–158

Lulka, D (2009) The residual humanism of hybridity: retaining a sense of the earth. *Transactions of the Institute of British Geographers* 34: 378–393

MacKellar, C and Jones, DA (2012) Introduction. In C MacKellar and DA Jones (eds) *Chimera's children: ethical, philosophical and religious perspectives on human-nonhuman experimentation*. London: Continuum, (no page numbers) Accessed 14 Jan 2019

McConnell, F (2017) Liminal geopolitics: the subjectivity and spatiality of diplomacy at the margins. *Transactions of the Institute of British Geographers* 42: 139–152

McConnell, F and Dittmer, J (2018) Liminality and British overseas territories. *Environment and Planning D: Society and Space* 36: 139–158

Mouffe, C (2005) *On the political*. Routledge: Abingdon

Owen, S, Moseley, M and Courtney, P (2007) Bridging the gap: an attempt to reconcile strategic planning and very local community-based planning in rural England. *Local Government Studies* 33 (1): 49–76

Parker, G (2012) *Neighbourhood planning: precursors, lessons and prospects*. 40th Joint Planning Law Conference, Oxford www.quadrilect.com/Gavin%20Parker.pdf. Accessed 15 Dec 2018

Parker, G (2017) The uneven geographies of neighbourhood planning in England. In S Brownill and Q Bradley (eds) *Localism and neighbourhood planning: power to the people?* Policy Press: Bristol, p75–91

Parker, G, Lynn, T and Wargent, M (2017) Contestation and conservatism in neighbourhood planning: reconciling agonism and collaboration? *Planning Theory and Practice* 18 (3): 446–465

Parker, G and Salter, K (2017) Taking stock of neighbourhood planning 2011–2016. *Planning Practice and Research* 32 (4): 478–490

Severi, C (2015) *The chimera principle*. University of Chicago Press: Chicago

Stephenson, J (2010) People and place. *Planning Theory and Practice* 11 (1): 9–21

Turner, V (1967) *The forest of symbols: aspects of Ndembu ritual*. Cornell University Press: Ithaca and London, p93–111

Turner, V (1969) *The ritual process: structure and anti-structure*. Aldine Transaction: New Brunswick and London

Turner, V (1982) Liminal to liminoid in play flow and ritual: an essay in comparative symbology. In V. Turner (ed) *From ritual to theatre: the human seriousness of play*. Performing Arts Journal Publications: New York, p20–60

van Gennep, A (1960) *The rites of passage*. University of Chicago Press: Chicago

Vigar, G, Gunn, S and Brookes, E (2017) Governing our neighbours: participation and conflict in neighbourhood planning. *The Town Planning Review* 88 (4): 423–442

Waite, D and Bristow, G (2018) Spaces of city-regionalism: conceptualising pluralism in policy making. *Environment and Planning C: Politics and Space* DOI:10.1177/2399654418791824

Wargent, M and Parker, G (2018) Re-imagining neighbourhood governance: the future of neighbourhood planning in England. *The Town Planning Review* 89 (4): 379–402

Whatmore, S (2002) *Hybrid geographies: natures, cultures, spaces*. Sage: London

Wills, J (2016) *Locating localism: statecraft, citizenship and democracy*. Policy Press: Bristol

Conclusion

Introduction

Throughout this book, I have engaged in detail with one specific neighbourhood planning process, in which I have been personally and deeply involved, to supplement the growing body of work addressing and assessing neighbourhood plans and planning collectively. My aims were multiple, and were informed by issues and questions identified in the extant literature, such as the need for more detailed information as to how the content of a neighbourhood plan is shaped and the microdynamics involved in doing so (Parker and Street, 2015; Brownill, 2017; Parker, 2017; Vigar et al, 2017), questions as to whether neighbourhood planning can resolve conflict between contradictory planning agendas (Bradley, 2015; Parker and Street, 2015; Brownill, 2017), and debates as to the extent to which neighbourhood planning really constitutes a significant transfer or reconfiguration of power (Gallent and Robinson, 2013; Wills, 2016a). Other concerns included whether and how neighbourhood planning and local planning might work more effectively together (Parker et al, 2015; Parker and Salter, 2017; Wargent and Parker, 2018), opportunities for opposition and resistance (Wills, 2016b; Etherington and Jones, 2017), and the prospects for neighbourhood planning to bolster active citizenship (Gallent and Robinson, 2013; Gunn et al, 2014; Wills, 2016a; Bradley and Brownill, 2017; Brownill and Bradley, 2017; Colomb, 2017; Parker, 2017; Parker and Salter, 2017; Vigar et al, 2017).

In the first two chapters, I set the scene for the substantive discussion to follow by providing brief background to the introduction of neighbourhood planning, the circumstances for my own neighbourhood planning experience in terms of the local government context and the development agenda, and the process, progress and general approach to the Wootton and St Helen Without Neighbourhood Plan. Subsequently, I turned my attention to conventional conceptualisations of space and place to lay the disciplinary groundwork for the interrogation of the constructions of place that followed. The distinction in both geographical and planning literatures between the abstract space of planning and the lived place of the neighbourhood fed into an exploration of performative and affective understandings of place, and stimulated questions as to whether normative assumptions that place attachment is a reliably positive motivating force that can be enhanced by neighbourhood planning hold in the face of doing neighbourhood planning. This signalled my attempt to move beyond concerns with metaphors of space, such as scale and network (Brownill and Bradley, 2017), to explore varied constructions of place as they are performed in the practice of neighbourhood planning.

In each of the subsequent chapters, I took my lead from a specific concern identified in the existing literature to guide my exploration of the construction of place and its implications for neighbourhood planning. In Chapter 3, my focus was on the dynamics and contradictions of place-making (Brownill, 2017), which I considered in the context of the framing of sustainable development within the NPPF, and the divergent understandings of the environment, the Green Belt and sustainable development between planners/developers and communities. I proposed that the neoliberal prioritisation of economic growth over the social and environmental pillars of sustainable development perceived in the NPPF actively sustains conflict between planning agendas because, even without the enforced compliance of neighbourhood planning, the NPPF effectively stacks the decks to ensure that development can always come out on top. This was found to be compounded by the different understandings of key terms, not just because planning favours formal, scientific definitions and metrics, but because those local understandings are established on the basis of lived experience in an area. It would be a mistake, though, to assume that this is just a matter of scale: it is a matter of personal connection to the specificities of a locality that are the result of phenomenological immersion in place but elided by the planning system and its emphasis on cumulative 'net gain'. Differences in the type of knowledge, the basis for that knowledge and the criteria against which development proposals are judged, all contribute to the occurrence and maintenance of conflict in planning, but I argued that the NPPF prevents political debate about the direction of development while the planning system excludes neighbourhood concerns and knowledge from consideration. Without addressing the divergent underlying understandings upon which place is constructed by the various parties in neighbourhood planning, and without allowing meaningful political debate, there seems little scope for the resolution of such conflict. Post-political attempts to depoliticise planning are ineffective: they submerge and sustain dissensus, preventing rather than allowing resolution, to the detriment of planning outcomes for the community.

Chapter 4 was prompted by concerns regarding the difficulties encountered by communities in articulating their needs and objectives in terms of strategy or principle (Gallent and Robinson, 2013), and examined our own attempts to do just that in relation to scalar, spatial and temporal aspects of strategy. We started out by questioning what strategic really means in the context of neighbourhood planning as a way of separating two objectives seemingly assumed by the LPA to be yoked together, to enable us to reflect community concerns about the Green Belt while supporting the development proposed, and then we established both a spatial strategy and a temporal strategy. Through a tiered spatial strategy, we articulated key community concerns about coalescence between settlements and the loss of specific green spaces within a strategic framework anchored on the character of our area. The temporal strategy aligned our work on the neighbourhood plan with the timetable for the emerging local plan, to maximise our chances of getting our plan made before the local plan could be adopted and to optimise our opportunities to nudge LPP2 into closer alignment with our neighbourhood plan. This chapter highlighted the potential to draw upon the perceived strategies of the local authority to support neighbourhood plan objectives, as in the case of pushing for full garden village principles to secure separation between the new development and surrounding settlements, as well as the strategic value of aligning the development of a neighbourhood plan with that of the relevant local plan. Although we were criticised for putting the cart before the horse in pushing on with our neighbourhood plan, we did so for strategic reasons and – irrespective of the outcome – it maximised our chances of delivering on our communities' objectives. It also suggests that neighbourhood planning can move beyond simply planning the niceties of a neighbourhood to influencing higher-level strategic policies, and I argued that such a step is necessary if neighbourhood planning is to deliver on its own stated aims.

In Chapter 5, my attention turned to performative constructions of place. The political construction of place was explored through the perceived attempt by the LPA to reimpose formal local government hierarchy following our decision not to delay the neighbourhood plan, in effect reshaping place as two parishes rather than one Designated Area. This, though, reinforced the unity between the parishes and reinvigorated those involved in the neighbourhood plan to push on even harder, thereby increasing rather than decreasing whatever risk the LPA might have perceived in our approach. Subsequently, the practical construction of place attended to softer forms of power that can be employed by LPAs under the guise of providing assistance to modulate policy wording and dilute community aspirations. While LPAs have diverse means at their disposal in this regard, communities can also employ tactics to resist such efforts or to apply their own pressure to the relative timescales for respective planning documents. Specifically, the temporal modulating effects of an LPA appearing to operate as if its local plan has already been approved was highlighted as a potential demotivating factor even if not a practical barrier to a neighbourhood plan, again indicating that how planning is performed in its specificities can impact on its substantive effectiveness. Finally, presentational constructions of place were considered in relation to the place presented in the neighbourhood plan itself. In this context the challenge of catering for different audiences was highlighted, which also applied in the public examination of LPP2, when the need to present the wrong (local, non-scientific) knowledge in the right (strategic, selective) language became apparent. This discussion highlighted the uneasy coexistence of two neighbourhoods within one neighbourhood plan, with the lived place of the neighbourhood partially encapsulated in the supporting text kept apart from the planned space of the neighbourhood delivered through the policies. It also signposted the importance of engaging with broader planning processes and higher planning powers in securing community gains, all of which reinforce the view that no meaningful power has been transferred to the people through neighbourhood planning.

Chapter 6 drew together the emergent analytical and critical threads from these chapters to consider the implications for theorising the spaces and places of neighbourhood planning, for the legitimacy of neighbourhood planning in its current form, and for the form, role and position of neighbourhood planning in the future. The potentially detrimental effects of neighbourhood planning on place attachment were explored in relation to hybrid, liminal and chimerical ideas of place, whereby the community's place attachment and the place of neighbourhood planning in the planning system can both be considered chimerical, in that the experience is one of radical fluctuation between extreme oppositional possibilities, which is both inconvenient in planning terms and unsettling in affective terms. By contrast, the space of the neighbourhood plan as a document was characterised as hybrid as it seeks to accommodate both extreme possibilities, while the place of the neighbourhood itself was characterised as variously all three (hybrid, liminal and chimerical) as the scope of possible outcomes on the ground is more many and varied than the binaries around which the neighbourhood plan is based. This notion of chimerical instability was identified as an incidental effect of our approach to neighbourhood planning but was proposed as amenable to development into both a tactic for use by communities to keep open unsettling possibilities that an LPA might rather close down, and a means by which a space of negotiation could be delimited in advance of planning negotiations getting underway to ensure that planning outcomes are more collaboratively established and gains and losses are more equitably shared.

Subsequently, the legitimacy of neighbourhood plans was considered in relation to consent, procedural (representative and participatory) and substantive forms. The evaluation of legitimacy was not positive, as consent was deemed to be questionable and uninformed at best, both representative and participatory forms of procedural legitimacy were deemed to be hindered by the stipulations of the NPPF, the power dynamics of the planning system, and the demands of neighbourhood planning, and substantive legitimacy was found wanting due to structural issues such as the need for conformity, issues of deliverability such as quantitative restrictions on affordable housing and environmental ambitions, inequalities and expectations. Finally, the prospects for neighbourhood planning were determined to be equally gloomy as even modifications to streamline and clarify the process will be inadequate in the absence of more significant changes to the planning system. It was proposed that the NPPF should be modified to establish a more level playing field between the different facets of sustainable development, and that the role of and relationship between neighbourhood and local plans should be reconfigured so that neighbourhood planning can function more strategically and local planning can be more sensitive to locality.

Significantly, the concerns discussed through these chapters raise the prospect of (neighbourhood) planning not just being post-political in shifting attention from substantive debate to technicalities but also post-democratic in simultaneously affirming and undermining democratic rights as egalitarian rhetoric masks increasing concentration of power in the hands of the economic elite (Crouch, 2004, 2016; Doucette and Kang, 2018). At the same time that residents and their local knowledge are invited into the planning process, that same process stifles, silences and excludes that knowledge: localism claims to democratise planning but the power remains in the hands of the already powerful. What is needed in the interests of enhancing the legitimacy of neighbourhood planning is the reintroduction of people into the political and the reinsertion of the political into planning so that strategic decisions are made in a more democratic fashion and the planning and delivery of development is more collaborative and equitable. To this end, it was suggested that we need to shift our emphasis away from different levels of planning document to focus more acutely on the stages of strategic planning from the more abstract or in-principle to the more emplaced and specific. To be clear, I am not arguing here that planning should not be plan-led (although consideration of alternative models might be fruitful) but simply that the content and role of different plans need not be restrictively proscribed as different plans can play different strategic roles depending upon local circumstances. It was also proposed that by developing the idea of chimerical instability into a mechanism for establishing the oppositional tensions relevant to specific neighbourhood and local planning practices it should be possible to delimit a space of negotiation that can avoid all the benefits of development going in one direction and all the disbenefits going in another, such that a more collaborative and equitable planning system can be crafted in which all parties give a bit and all parties gain a bit.

If such modifications can be instantiated, then the detrimental impacts of neighbourhood planning on place attachment and the epistemic violence wreaked on communities and their local knowledge (Parker et al, 2017; Bradley, 2018) might largely be avoided as the space of negotiation would preclude the very worst instances from arising and the mutually agreed distribution of planning activities between local and neighbourhood plans can be tailored to accommodate local priorities, delivering a more explicitly political and more democratic planning process. Without such systemic alterations, however, it seems unlikely that neighbourhood planning will ever deliver on its potential or that it will ever gain legitimacy. Worse, it could lead to the progressive and terminal erosion of the very place identity and attachment upon which it depends and which it claims to validate. Further, it could erode civic participation in local democratic processes beyond repair due to the opacity of what is involved and the seeming impossibility of securing substantive gains for the community.

WSHWNP update

Between drafting the bulk of the manuscript immediately after the neighbourhood plan was submitted to the LPA and the submission deadline for the manuscript of this book, a full seven months have elapsed. At the time of submission, consultation is underway on the main modifications made to LPP2 following its public examination, while the neighbourhood plan is undergoing its own public examination. Although the final outcomes are still unknown, the early stages of these two examination processes give an indication of some of the likely outcomes for our neighbourhood, so this section considers what we might and might not achieve, and why.

It is incredibly difficult to pinpoint specific achievements that are directly attributable to the neighbourhood plan. Certain changes to the proposals for the Strategic Development Site might suggest that we have had some influence, as they support community objectives, but in each case there is a qualifying perspective that questions this supposed influence. For example, the removal of the fields between Whitecross and the proposed development from the strategic allocation following the first round of consultation on LPP2 suggests that our consultation responses pushing for precisely this change had been considered and deemed appropriate. However, during that same consultation process, the DIO's own proposals for their land identified those fields as constituting a strategic gap and removed these fields from consideration for development. Given that the DIO is a major stakeholder in the development, it is far more likely that the removal of these fields from the development proposal was in recognition of the DIO's consultation response rather than our own. Similarly, the initial finding of the examination of LPP2 that neither the playing fields at Dalton Barracks nor an area immediately adjacent to Whitecross (both of which we sought to designate as Local Green Space), should be removed from the Green Belt and therefore should not form part of the development proposal, suggests that we had been influential in this regard. However, no reference was made to the neighbourhood plan in this finding so these determinations might have been made for entirely different reasons irrespective of our views. Consequently, even if we have had some degree of influence, we cannot consider it a success as we do not know that this is the case.

On the one hand, our perspective seemed to receive some consideration as we were asked to submit our plan to the examination of LPP2 and the LPA was encouraged to reconsider and liaise with us on certain issues. On the other hand, the modified version of LPP2 currently undergoing consultation no longer describes its proposal for our area as a new settlement but simply as development. These seemingly contradictory indicators primed us nicely for further experience of chimerical instability, sustaining uncertainty and erraticism in place identity and attachment as sentiments flicker between anticipation verging on elation and frustration sliding into despondency. More significantly, though, given our concerns about the lack of separation for Shippon, is the way in which the change in emphasis from settlement to development further reinforces the intention to merge the new development with Shippon, which positions LPP2 and the LPA even further away from the WSHWNP and the local community, further entrenching rather than resolving the conflict between the parties. While there remains some hope that we might still influence the outcome and while both the receptivity of the examination towards our attempted reconfiguration of the structural relationship between the local plan and the neighbourhood plan, and the temporal relationship between them in terms of which will be completed first, remain uncertain, I am not hopeful for a happy outcome for the community. Consequently, although this evaluation remains somewhat speculative, it increasingly seems to me that we will have failed to achieve our objectives either through direct participation in LPP2 or through our own strategic approach to neighbourhood planning. This is both deeply disappointing and hugely frustrating given that it is entirely feasible to deliver both community and strategic objectives and it is inexplicable to me why such an outcome would not automatically be pursued as a primary objective where such mutual benefits can be secured, especially given the stated aims of localism. The apparent impossibility of securing community gains even when doing so would pose no impediment whatsoever to development objectives highlights yet again the unassailability of dominant agendas, vested interests and institutionalised power dynamics, which further undermines the claims of localism as a transfer of power to the people.

Ultimately, then, we are left in a position whereby we can be fairly certain that we have failed to achieve the primary objectives of our community despite our sustained, varied, strategic and considerable efforts, but we cannot be at all certain that we have had any influence in generating results that do deliver beneficial outcomes for our communities. This has dire implications for the legitimacy of neighbourhood planning. It raises serious questions about the role of neighbourhood planning in adding value to the planning system by securing a community voice (Parker et al, 2019) if that voice is prevented from doing anything other than shout into the wind, undermining procedural legitimacy. It also suggests that it is extremely difficult to evaluate what difference, if any, a neighbourhood plan might have made, making it impossible to establish substantive legitimacy. In terms of participatory democracy, further doubts are raised as it is unlikely that many volunteers will remain involved if they can only ever be certain of their failures but never gain the satisfaction of successes, even if or where there are some. Not only is the neighbourhood planning process punishing in its enactment, but the outcome of that process appears to be a classic case of no good deed going unpunished.

These issues are compounded by the proceduralised eviction of an NPSG from the closing stages of the neighbourhood plan and the progressive colonisation of the process by an LPA. While it makes sense for an LPA to take responsibility for the organisation of the referendum, the channelling of all correspondence between an examiner and a community through an LPA is questionable. In many instances, this would probably not be problematic but in circumstances where the relationship between a community and an LPA has broken down or is difficult, this raises another concerning imbalance in power. While an LPA can see the community's responses to the clarification questions posed by an examiner, possibly even change them and in any event formulate counterarguments to them before forwarding them to the examiner, the community does not have a similar opportunity to pre-empt the responses of their LPA. This is further compounded by the non-binding nature of the examiner's report (Parker et al, 2019). Even if an examiner supports or upholds community objectives, if an LPA is not obliged to accept their recommendations, then seemingly the examination outcome can simply be ignored. This might be unlikely due to the requirement for further consultation and potentially a second examination, but even this eventuality could be used by an LPA seeking to delay a neighbourhood plan. Moreover, the difficulties do not end there, as it seems that a qualifying body can only withdraw a neighbourhood plan in response to the examiner's report and not to what their LPA decides to do with that report (NPIERS, 2018). If this is the case, where an NPSG is happy with the changes recommended by the examiner but their LPA decides to modify the neighbourhood plan in a manner that diverges from those recommended changes, the NPSG is seemingly powerless because the opportunity for them to withdraw their plan has already passed by the time they are informed of their LPA's intentions. Yet again, the voice of the community is silenced, and power is returned to the LPA. Little wonder, then, that in extreme cases an NPSG is left with no choice but to campaign against their own neighbourhood plan because it is no longer *their* neighbourhood plan (see Milne, 2016) in terms of either its substantive content or its practical development, and the legitimacy of neighbourhood planning falls even further, if that is even possible.

The extent of this progressive eviction of the community from the development of its own plan is indicated by the exhortation that communities should seek to communicate with the LPA in these final stages to ensure that 'as far as possible' it continues to reflect community aspirations and retain community ownership (Parker et al, 2019: 93). This is interesting wording as it suggests that 'as far as possible' is not very far at all, further undermining both the spirit and potential of the Localism Act. It also renders questionable the assertion that the important thing about neighbourhood planning is not the process but the substance of the plan (Parker et al, 2019), as – on the one hand – the final substance of the plan is inextricably bound to the process of its development, and – on the other hand – it appears that the community is progressively and forcibly excavated from both the process and the substance of its own neighbourhood plan. The outcome that seems most likely in this scenario is that the one thing that a neighbourhood plan neither reflects nor respects is the neighbourhood to which it is meant to relate and from which it originally sprang. While this is certainly important in terms of the substance of a neighbourhood plan, it is perhaps even more important in terms of the process as it is the process that stipulates and ensures the progressive eviction of the community from the development of its own plan, which in turn has implications for its substance. From my perspective, the important thing about neighbourhood planning absolutely *is* the process – and specifically the deficiencies inherent to it – as without the process there would be no substance.

This progressive sidelining of those involved at just the stage in the process when any achievements in terms of community gains might start to become clear, is also baffling in terms of both the legitimacy of neighbourhood planning and its supposed role in enhancing participatory democracy. To my mind, this shift in responsibility is understated in both the regulations and guidance on neighbourhood planning, further undermining the legitimacy of the process as it impedes the capacity of volunteers to give informed consent if they do not know what it is to which they are consenting or when their involvement will cease. Equally, slaving away for years on a document that you are then disbarred from seeing through to completion is galling as it denies and devalues the time, effort and emotion invested in the neighbourhood plan by those involved, which is hardly conducive to encouraging ongoing or repeated participation. Consequently, I would argue that this enforced eviction of the community from the final stages of plan development is yet another damning indictment of neighbourhood planning as a form of localism, as an effective planning vehicle and as a step towards participatory democracy.

Thematic insights and proposals

No doubt those in the development industry and planning profession, and those who are advocates of neighbourhood planning will think I am being overly cynical, possibly disingenuous, perhaps even inflammatory. That is not my intention (or at least it is only my intention in so far as it is a rhetorical strategy), but reflecting upon my own experience of neighbourhood planning in the context of the academic literature does highlight numerous insights that have relevance far beyond the specific case study from which they arose, with potential to inform both practical and critical engagements with varied forms of localism in diverse socio-cultural and political settings. As these insights – in a planning sense – nest with one another, they are addressed here in succession from opportunities arising within neighbourhood planning to opportunities to overhaul the broader planning project, and they are broadly characterised around thematic issues of power, structure, relationality, context, creativity, trajectory, critique, chimericality and place.

Power

The existing literature encourages the exploration of and experimentation with different ways of establishing opposition from within the existing planning system and hierarchy (Parker and Street, 2015; Parker and Salter, 2017), and our experience shows that this can indeed be fruitful. In practical terms, our questioning of what constitutes strategy in a neighbourhood planning context not only targeted debate at issues that normally would not be considered but also enabled us to voice dissent within our neighbourhood plan, even if only in the supporting text seemingly considered irrelevant for planning purposes. Critically though, such an approach on its own is fundamentally self-limiting as it does nothing to address the excessive constraints on what neighbourhood plans can or should do or the power dynamics at play and the varied forms in which power takes shape. Such an approach simply accepts the status quo and is therefore too dependent upon both individual and institutional differences between communities and LPAs. Consequently, we need to start thinking more creatively about . . .

Structure

The existing literature acknowledges that there is potential to unsettle the attempts at policy modulation or to work outside the formal space of neighbourhood planning (Parker and Street, 2015; Wills, 2016b), and our experience shows that this is entirely possible. Our resistance to proposed wording changes, our resistance to perceived attempts to delay our progress and our engagement with the examination of LPP2 all demonstrate our attempts to unsettle or work around the formalities of neighbourhood planning. Critically, though, this also does not overtly challenge the formal structures of power and operating outside one's allocated structures and systems is a risky business as there is an ever-present risk of being put back in one's box. Such an approach is too dependent on planning inspectors, extending the hierarchical power to which neighbourhood planning is subjected, thereby further undermining the transfer of power to the people, especially as the potential influence of an independent examiner is constrained by the non-binding nature of their recommendations. The consequence of this is that such approaches are destined to remain the exception rather than the rule as it makes no attempt to mainstream this way of working. If we are to engage in localist initiatives, we should be doing so in a localist spirit rather than relying on the very systems and structures that we seek to unsettle in order to achieve our goal. However, the intertextual world of planning means that in order to do this we must be mindful of . . .

Relationality

There are growing calls in the literature for reconsideration of the relationship between local planning and neighbourhood planning (Parker and Salter, 2017; Parker et al, 2017; Wargent and Parker, 2018), and our engagement with neighbourhood planning sought to disrupt the conventional hierarchy in practice. We raised challenges to the scalar assumption that strategy can only be done at the level of the local, and we employed our own spatial and temporal strategies to maximise outcomes for our community. I fully concur that we need to reconsider this relationship, and I suggested ways in which it might be reconfigured, such that the emphasis shifts away from a document hierarchy towards a strategic process so that local planning can become more sensitive to locality and neighbourhood planning can contribute to strategy. I would also argue that where both community and strategic objectives can be delivered simultaneously, this should be the overriding aim and a prescribed duty, rendering both sets of objectives equivalent and helping to resolve conflict rather than the current reinforcement of conflict through the stipulated primacy of local plan strategic objectives that inevitably outrank community objectives even where the two are entirely compatible. The critical point here, though, is that while I would argue that these are necessary steps, they are not sufficient, because not only are the documents of planning inter-related but some still have more power than others, so we also need to consider the broader . . .

Context

The NPPF is available as a strategic resource for neighbourhood planning teams in so far as they can refer to it in support of their community objectives, as we sought to do in relation to the Green Belt and sustainable development. Critically, though, the key point here is that concerns over the need for neighbourhood plans to conform with a local plan and the development allocated within it risks missing the broader point that deleting this need for conformity would do nothing on its own to redress the prioritisation of economic growth identified in and enforced through the NPPF. As such, it is not simply neighbourhood planning that needs to be revised, or its relationship with local planning, but the NPPF, as only then can any changes to neighbourhood planning as a specific instantiation of localism bring about tangible changes for communities. Unfortunately, this still does nothing to address the persistent exclusion of local knowledge, place attachment and community concerns from planning deliberations by the vested powers and professional conventions that govern how planning operates. Addressing these deficiencies demands . . .

Creativity

The potential and need for creativity and innovation within neighbourhood planning is acknowledged in the literature (Parker and Street, 2015; Parker and Salter, 2017) but less clearly or commonly articulated is the potential and need for creativity in the reformulation of neighbourhood planning. Recognising the need to establish firmly the role of citizen-planners in the wider system (Parker et al, 2019) is one thing, but we also need to ask probing questions about what that role could and should be, without relying on patronising and constraining assumptions about the incapability of community volunteers to think strategically. Practically speaking, our neighbourhood plan sought to reformulate what a neighbourhood plan might do and how it might relate to a local plan, but from a critical perspective, this was done entirely from within the practice of neighbourhood planning so was itself inherently constrained in its perspective. Applying this to planning more broadly, perhaps the habitus of planning as much as the formal prescriptions of the planning system is mitigating against more creative consideration of the potential of neighbourhood planning (Stephenson, 2010). If localism is about bringing into the formal planning system local, unprofessional knowledge and alternative perspectives, then why should this not be applied to the strategic matters from which communities are currently excluded or indeed the ideological project of planning within which they are now increasingly enrolled? While bringing outsiders into planning might hold potential for the creative reworking of neighbourhood planning, it also holds potential to impact on the very place attachment upon which neighbourhood planning depends, which means that we need to be mindful of . . .

Trajectory

The positive affective power of place attachment and the potential for neighbourhood planning to enhance such sentiments is recognised in the literature (Bradley, 2017; Vigar et al, 2017; Lennon and Moore, 2018), but the potential for place attachment to suffer as a result of participation in neighbourhood planning is conspicuous by its absence. Place attachment is fluid and contingent, and I suggest it becomes especially vulnerable in the context of neighbourhood planning, specifically due to the frustrations of the process, the uncertainty of its outcomes, the exclusion of local knowledge, the dismissal of local concerns and the perceived erosion of the very qualities of a place that generate place attachment in the first place. We therefore need to exercise caution in our utilisation of place attachment as a resource for neighbourhood planning as experiencing a place as chimerically unstable is distinctly unpleasant and neighbourhood planning risks destroying that which it claims to nurture, with long-term implications for participatory democracy. Consequently, there is arguably a strong and growing need for effective localist . . .

Critique

The existing literature on neighbourhood planning is replete with concerns for reforming at least some of its elements, for example, to make the process less onerous, to clarify the relationship between the community and the LPA or to reduce the need for policy wording to be amended (Parker et al, 2015, 2017; Vigar et al, 2017; Wargent and Parker, 2018). However, none of the literature to my knowledge goes so far as to call for an explicitly localist strand of critical scholarship. This volume supports and extends these concerns, indicating that there are major failings in neighbourhood planning as it is currently configured, and in fact that these failings go beyond neighbourhood planning per se. Such shortcomings not only mean that neighbourhood planning cannot deliver on its aims, but also that it fundamentally fails in terms of legitimacy. This is not so much due to the commonly reported issues of self-selection and lack of representativeness (Cowie and Davoudi, 2015; Gunn et al, 2014; Wills, 2016a, 2016b; Parker, 2017) – although these are cause for concern – but because on every dimension explored (consent, procedural [representative and participatory] and substantive) the very initiative and process that claim to validate local knowledge and community perspectives have been shown to deny, exclude, dismiss or otherwise delegitimise them in a systematic and sustained manner. Add to this the potentially harmful effects of participation in neighbourhood planning on the very place attachment that underpins and feeds it, and the picture looks very bleak indeed. Whether an incidental effect or an intentional design matters not: localities risk being progressively dismantled. From my experience of neighbourhood planning, local knowledge claims to be validated but is denied, community participation is invited but dismissed, place attachment is purportedly valued but potentially destroyed. An explicitly localist strand of critical scholarship could challenge these trends in an integrative, sustained, evidence-based and co-ordinated fashion. At the same time, and acknowledging that critiquing events is easy but providing solutions is more difficult, it is also perhaps time to explore the productive potential within . . .

Chimericality

It is bitter-sweet, then, that the potentially detrimental effects of neighbourhood planning on place attachment might yet provide part of the solution. The preceding paragraphs emphasised the interconnected nature of the various shortcomings and deficiencies of neighbourhood planning within the context of the broader planning project, and summarised the proposals articulated in this volume to reconfigure planning so that planning could be more political, neighbourhood planning could be more strategic, and local planning could be more sensitive to locality. Left outstanding, though, is the urgent need to redress the power inequalities among the varied actors in the planning system, as without a more equitable distribution of power the community will always come off worst. In this regard, chimerical instability can play two roles. Within an un-reconfigured neighbourhood planning system, chimerical instability can be introduced into the planning hierarchy by neighbourhood planning teams thinking and acting strategically to unsettle the conventional roles, relations and assumptions within that hierarchy. In essence, this is a means of prompting professional planners to shake themselves free of the stipulations and habitus of their profession in the hope that more creative solutions to planning dilemmas might be sourced from outside the conventional boundaries of planning. Alternatively and additionally, chimerical instability can be developed into a conceptual and methodological tool to identify the axes of tension and extreme oppositional possibilities at play in the planning process as a means of preventing those extremes from materialising and to establish a space of negotiation with potential to deliver more collaborative planning deliberations and more equitable planning outcomes. This, of course, brings us nicely to the final theme of:

Place

Alongside a need to think more creatively about the spatialities of (neighbourhood) planning, we also need to engage more creatively with the places that are neighbourhoods. It is too simplistic to rely on crude distinctions between the lived place of the neighbourhood and the abstract space of planning that deny the capacity of local people to think with and experience space abstractly and deny the capacity of planners to dwell in place. We need to bring the phenomenological and the abstract together in the planning process rather than holding them apart, by shifting the effort away from stipulating development through formal plans and policies that are foisted upon communities without consent to envisioning a future jointly with communities that is then crafted into a delivery plan. This, of course, returns us to the need for communities to be involved in strategic aspects of planning, but it also prompts a shift away from table-top consultation exercises that constrain planning activities to the already-abstracted form of plans towards more place-based engagement activities that facilitate on-the-ground envisioning of the future and translation of that vision into agreed plans. This would be not only more collaborative but also more egalitarian in the forms of knowledge that are brought to bear on planning, underlining both the importance of planning in directing development and the significance of the particularities of place for communities, thereby enhancing both the effectiveness and perceived legitimacy of planning.

Some form of modification along these lines is – to my mind – a necessity. In our case, I see absolutely no reason why both community objectives of settlement separation and LPA objectives for the Strategic Development Site could not be delivered by establishing the new development as a stand-alone settlement. Yet it seems that we are destined not to secure our objective despite the claims of the localist agenda, despite our sustained and robust involvement in every stage of the consultation process for the local plan, despite our own strategic approach to neighbourhood planning and despite the additional weight acquired by our own plan as it progressed alongside the emergence of the local plan. Even if meaningful separation is eventually provided, we would not be able to claim the credit for that as it would be dependent upon the decisions of national level planning inspectors. Let me be clear here: I am not for one moment suggesting that there is anything untoward or illicit about the actions of the LPA in our situation. They have been acting in accordance with the neighbourhood planning process as appropriate for their own role and agenda just as we have been doing in accordance with ours. However, that is precisely the critical point at stake here: it is the system and process that are the target of my critique, not the actions of specific individuals and organisations involved. My evaluation, based upon my own experience, is that the very architecture and process for neighbourhood planning – set up as it is within the broader planning framework in these neoliberal, post-political and post-democratic times – systematically mitigate against the achievement of substantive outcomes for the community. This is not to suggest that such outcomes are impossible to achieve but to assert that the odds are stacked against the community to such an extent that communities are unlikely to make any substantive gains, especially if they impact – however minimally – on the other interests involved, unless somebody in authority further up the planning hierarchy is especially enlightened with respect to localism. If consensual, collaborative and equitable outcomes are not delivered by a supposedly localist planning system even when it is entirely possible to do so, what hope is there, really, for communities? If even communities who are supportive of development, engage in planning in good faith, and request only minimal concessions through their legitimate local and neighbourhood planning involvement are still denied substantive recognition or accommodation, is it any wonder that the lack of faith in statutory bodies and systems on the part of citizens is as corrosive as my own sentiments evidence, and will it be any wonder if participation rates tumble and disaffection soars? It is acknowledged that neighbourhood planning was introduced to reduce public opposition to development in the short term (Gallent and Robinson, 2013; Bradley et al, 2017), but the long-term implications and motivations are given less consideration. Given the concerns raised in these chapters, it seems at least possible that its long-term post-democratic outcome might be the generalised demolition of public engagement and participation through the progressive evacuation of the public from the politics of planning locally by enrolling citizens in planning processes that systematically engender disaffection with the political process as a whole. While I'm not suggesting that such an extreme outcome is either an intentional goal or co-ordinated strategy of the state, such an outcome is a worrying prospect even as an incidental by-product of a well-meaning initiative, and under circumstances such as those discussed in this volume, I cannot consider neighbourhood planning to be anything other than farcical, regressive and duplicitous.

Conclusion

Consequently, I fully concur with assertions that there is something deeply regressive and deeply troubling about the way in which localism – especially neighbourhood planning – is being enacted, and that the issues identified here and elsewhere should be raising far more public concern than they are (Tomaney, 2016). However, to my mind, they should also be raising far more academic concern than appears to be the case and proposals for the future should be far more creative and radical than they have been to date. If a state-led project of localism cannot deliver localism in any meaningful sense through its own project then it cannot make sense to put the responsibility for doing localism more creatively on the shoulders of community volunteers embroiled in the process, who find themselves bound not only by the pre-existing power dynamics and structural determinations of planning but also by the newly imposed constraints of the supposedly emancipatory localist project. At the very least, we need a more co-ordinated branch of explicitly localist critical scholarship, and preferably, we need a radical review and revision of neighbourhood planning in its broader planning framework.

This must surely be the case where a neighbourhood plan's only chance of delivering the objectives of its community is by being and becoming that which it is not permitted to be. We needed to be strategic when we were not meant to be; we needed to complete our plan first when we were meant to come last; we needed to act as if we were professional experts when we were not. The latent power and potential within neighbourhood planning to which this speaks is reinforced by a seemingly similar manoeuvre perceived on the part of the LPA in acting as if its local plan had already been approved when it had not. In this light, the future for neighbourhood planning should be bright, but it is not because the potential of neighbourhood planning is quashed on every front to such an extent that any revision of neighbourhood planning in isolation would achieve nothing due to the inter-textual nature of planning, the power dynamics at play, the ideological framing of sustainable development in the NPPF, and the false distinctions that are maintained between the different levels of planning which sustain focus on discrete entities and their respective exclusions in the abstract rather than complementary processes and their integrative outcomes on the ground. Consequently, I offer three concluding recommendations. We need to stop policing the gaps between our respective roles and plans and start pooling our resources among them, dwelling in their overlaps and contiguities, and recognising the capacity of communities to think and act strategically and the need for planning to be more sensitive to locality and community. We need to delimit a framework for planning and a space for negotiation that ensures a more equitable distribution of gives and takes to ensure that community concerns are meaningfully accommodated. We need to take the implications of neighbourhood planning for place identity and attachment more seriously, and we should work strategically with such conceptualisations both as a tactic of opposition within neighbourhood planning and as a tool for the reformulation of neighbourhood planning. By such means, we might pave the way for the generation of more creative, collaborative and equitable solutions to planning dilemmas and conflicted agendas, and perhaps then neighbourhood planning will be able to escape my accusations of ineffectiveness and illegitimacy.

References

Bradley, Q (2015) The political identities of neighbourhood planning in England. *Space and Polity* 19 (2): 97–109

Bradley, Q (2017) Neighbourhood planning and the impact of place identity on housing development in England. *Planning Theory and Practice* 18 (2): 233–248

Bradley, Q (2018) Neighbourhood planning and the production of spatial knowledge. *The Town Planning Review* 89 (1): 23–42

Bradley, Q and Brownill, S (2017) Reflections of neighbourhood planning: towards a progressive localism. In S Brownill and Q Bradley (eds) *Localism and neighbourhood planning: power to the people?* Policy Press: Bristol, p251–267

Bradley, Q, Burnett, A and Sparling, W (2017) Neighbourhood planning and the spatial practices of localism. In S Brownill and Q Bradley (eds) *Localism and neighbourhood planning: power to the people?* Policy Press: Bristol, p57–74

Brownill, S (2017) Assembling neighbourhoods: topologies of power and the reshaping of planning. In S Brownill and Q Bradley (eds) *Localism and neighbourhood planning: power to the people?* Policy Press: Bristol, p145–161

Brownill, S and Bradley, Q (2017) Introduction. In S Brownill and Q Bradley (eds) *Localism and neighbourhood planning: power to the people?* Policy Press: Bristol, p1–15

Colomb, C (2017) Participation and conflict in the formation of neighbourhood areas and forums in 'super-diverse' cities. In S Brownill and Q Bradley (eds) *Localism and neighbourhood planning: power to the people?* Policy Press: Bristol, p127–144

Cowie, P and Davoudi, S (2015) Is small really beautiful? The legitimacy of neighbourhood planning. In S Davoudi and A Madanipour (eds) *Reconsidering localism.* Routledge: New York and London, chapter 9. (no page numbers) Accessed 6 Dec 2018

Crouch, C (2004) *Post-democracy.* Polity: Cambridge

Crouch, C (2016) The march towards post-democracy, ten years on. *The Political Quarterly* 87: 71–75

Doucette, J and Kang, S (2018) Legal geographies of labour and postdemocracy: reinforcing non-standard work in South Korea. *Transactions of the Institute of British Geographers* 43 (2): 200–214

Etherington, D and Jones, M (2017) Re-stating the post-political: depoliticization, social inequalities, and city-region growth. *Environment and Planning A: Economy and Space* 50 (1): 51–72

Gallent, N and Robinson, S (2013) *Neighbourhood planning: communities, networks and governance.* Policy Press: Bristol

Gunn, S, Brooks, E and Vigar, G (2014) The community's capacity to plan: the disproportionate requirements of the new English neighbourhood planning initiative. In S Davoudi and A Madanipour (eds) *Reconsidering localism.* Routledge: New York and London, chapter 8. (no page numbers) Accessed 6 Dec 2018

Lennon, M and Moore, D (2018) Planning, 'politics' and the production of space: the formulation and application of a framework for examining the micropolitics of community place-making. *Journal of Environmental Policy and Planning* DOI:10.1080/1523908X.2018.1508336

Milne, R (2016) Residents reject Derbyshire neighbourhood plan. *Planning Portal* www.planningportal.co.uk, news article 434 Accessed 10 Jan 2019

NPIERS (2018) *Guidance to service users and examiners*. Neighbourhood Planning Independent Examiner Referral Service

Parker, G (2017) The uneven geographies of neighbourhood planning in England. In S Brownill and Q Bradley (eds) *Localism and neighbourhood planning: power to the people?* Policy Press: Bristol, p75–91

Parker, G, Lynn, T and Wargent, M (2015) Sticking to the script? The co-production of neighbourhood planning in England. *The Town Planning Review* 86 (5): 519–536

Parker, G, Lynn, T and Wargent, M (2017) Contestation and conservatism in neighbourhood planning: reconciling agonism and collaboration? *Planning Theory and Practice* 18 (3): 446–465

Parker, G and Salter, K (2017) Taking stock of neighbourhood planning 2011–2016. *Planning Practice and Research* 32 (4): 478–490

Parker, G, Salter, K and Wargent, M (2019) *Neighbourhood planning in practice*. Lund Humphries: London

Parker, G and Street, E (2015) Planning at the neighbourhood scale: localism, dialogical politics, and the modulation of community action. *Environment and Planning C: Government and Policy* 33: 794–810

Stephenson, J (2010) People and place. *Planning Theory and Practice* 11 (1): 9–21

Tomaney, J (2016) Limits of devolution: localism, economics and post-democracy. *The Political Quarterly* 87 (4): 546–552

Vigar, G, Gunn, S and Brookes, E (2017) Governing our neighbours: participation and conflict in neighbourhood planning. *The Town Planning Review* 88 (4): 423–442

Wargent, M and Parker, G (2018) Re-imagining neighbourhood governance: the future of neighbourhood planning in England. *The Town Planning Review* 89 (4): 379–402

Wills, J (2016a) *Locating localism: statecraft, citizenship and democracy*. Policy Press: Bristol

Wills, J (2016b) Emerging geographies of English localism: the case of neighbourhood planning. *Political Geography* 53: 43–53

Index

agonism/antagonism 27, 59, 64–65, 67, 89, 118

Basic Conditions Statement 24–25, 94, 113
binary/ies 31, 38, 104–105, 107–109, 119
brownfield 71–72
buffer/s 78–79, 99

Character Assessment 25, 75–76, 88
chimerical/ity/ies 106–110, 119, 122, 128, 136–137; instability/ies 104–106, 109–110, 111, 119, 128–130
chimerically unstable 109, 119, 136
community/ies 39, 51, 88–92, 103–104, 130–132; Infrastructure Levy (CIL) 15–16, 73; -led/-based/-grounded 1–3, 15–18, 70, 89, 91, 95–96, 106; objectives 51, 60, 92, 130–131, 134, 137
consultation 23–28, 58–59, 75–80, 130, 137–138; Statement 24–25, 53, 57, 60

Dalton Barracks 20, 26–27, 59, 70, 79, 90
democracy/atic 18, 95, 98, 104, 106, 111, 113, 118–119, 129; democratise/ising/isation 31, 65, 128; participatory 9, 111–113, 131–132; post-democratic 128, 138
designated area 23–25, 35, 55, 69, 88, 113
duty: to cooperate 6, 20–22, 27; to support 8, 45, 89, 97, 122

economic growth/development 9, 15, 19–20, 48, 59, 119
economy first 9, 50–53, 63, 72, 114, 118
environment/al 15–17, 49–56, 76–77; strategic environmental assessment (*see under* strategic)
examination 24–28, 71–74, 80–83, 94, 98, 129–131

exceptional circumstances 19, 58, 71
expertise 39, 53, 64, 93–94

garden village/city principles 76–77, 90, 97–98
garden village/s or community/ies 76–77, 90
governance 14, 33–34; local 13–14, 62
Green Belt 51–52, 56–60, 70–73, 87–88; study 57–62

Habitats Regulations Assessment (HRA) 24, 113
heritage assets 15, 23, 26
hybrid/ity/ies 104–109, 119, 128

institutional/ised 115, 118, 131; oversight 87, 120; parochialism 63

knowledge/s 54, 64, 90, 93, 96, 98, 126; everyday/lay 39, 42, 52–56, 60, 95, 106, 121; formal/informal 41, 44, 95, 52–56, 62; local/place 31, 39–42, 65, 96, 135–136; professional/scientific/technical 39, 41, 52–56, 60, 68, 93

legitimate/acy 91, 110–115, 128–129, 131–132; framework of 111–115, 121
legitimise/ation/de-/re- 52, 56, 64–65, 110, 136
liminal/ity/ies 105–109, 119
local 112–114, 116–118; Green Space (LGS) 15, 24, 75–76, 78, 99; plan/s 32, 80, 121; planning authority (LPA) 28, 44, 71, 77–78, 89–98, 117, 120, 131–132
localism 13–16, 32–34, 114–118, 128, 131, 139; Act 13, 20, 52, 65
localist critique 136

LPP1/Local Plan Part 1 26, 70, 74, 79–80
LPP2/Local Plan Part 2 20–22, 27–28, 70–72, 78–82, 93, 129–130

multi-naturalism 64

National Planning Policy Framework (NPPF) 48–52, 56–57, 70–71, 111–116, 121–122
neighbourhood 106–109; plans 15–17, 32, 51, 79–84, 97, 120–122; planning 14–18, 23–24, 31–43, 79–81, 87–100, 110–115; planning regulations 21–23, 69, 74, 78, 92, 121, 132
NIMBY/ism/ist 9, 58–59, 63, 81, 83, 103
non-representational 43–44

Oxford/shire 19–22, 33–34, 53, 64–65, 71, 80, 72; Growth Board 19

parish council/s 23–24, 33, 87–90, 113
place 38–44, 104–110, 137; attachment 38–43, 103–110, 135–136; knowledge 31, 39–41, 44–45, 63; sense of 39–44, 73, 84
planning 89–91, 95–96, 115–120, 133–138; collaborative 41, 64–65, 116–120, 122; hierarchy 33, 79, 99, 137; system/framework 32–35, 48–49, 81–83, 110, 117, 121
politics/political 27, 48, 64, 87–92, 129; politicisation/de-/re- 51, 64, 89, 126; post-political 27, 51, 89, 126, 128, 138
power 38, 52, 86–87, 98–99, 111, 115, 133; dynamics 18, 32, 43, 139; formal 49, 87–91, 120–121; inequalities 28, 86, 113–114, 137; soft 91–95, 114; transfer/reconfiguration of 32, 94, 99, 127, 131, 134

referendum 24–25, 96–97, 113, 121

scale/scalar 34–36, 53–55, 59, 62, 69–73
Screening Opinion 24–25, 96
settlement hierarchy 56, 59, 74, 77
Shippon 26–27, 40, 70, 74, 76–78, 130
Site of Special Scientific Interest (SSSI) 50
social 39–40, 49–51; equity/justice 50, 62, 98, 110; inclusion 23, 87, 115
society 14, 17, 60, 64; Big Society 14
space 31–39, 109; of negotiation 118–120, 128–129; of plan/s or planning 43, 88–91, 95–100, 104–110; shared symbolic space 7, 49, 60, 64–65
spatial 7, 32–36, 73–78, 88; concepts 31–45, 105–109 (affective 38–43, 104–105, 135; chimerical (*see under* chimerical/ity/ies); containerised 32–35; horizontal 31–35, 88; hybrid (*see under* hybrid/ity/ies); liminal (*see under* liminal/ity/ies); networked 33–35; performative 40–44, 86–100, 113–114; practical 40–44; vertical 31–35, 88); experience 43–44, 103, 109, 119; imaginaries 6, 32–37; strategy (*see under* strategy)
spatiality/ies/isation 33, 36, 42, 62, 91, 109, 137
statutory 23, 26, 116, 120–122; body/ies 23, 34, 62, 81, 83; period for consultation 24; power/status/weight 15, 52, 95, 120
St Helen Without 22, 35
strategic 69–73, 78, 100, 117, 121, 128, 137–139; development site/allocation 16–17, 26, 59, 69–70, 73–77, 129; Environmental Assessment (SEA) 24, 53–54, 94; Green Gaps 25, 75, 80, 99; objectives 3–4, 69, 72–73, 97, 130, 134; thinking 9, 68, 73–74, 79, 82–83, 117, 135
strategy 18, 68–69, 74, 83, 99, 115, 138; spatial 25, 61, 69, 73–78, 99; temporal 69, 74, 82, 88, 126
sustainability 16, 60–63, 115; appraisal 24; trap 61
sustainable 15, 59, 61, 72–74; development 48–52, 60–63, 76, 115, 106, 116

temporal strategy *see under* strategy
temporality/ies 42, 54

understanding/s *see under* knowledge/s
unmet (housing) need 20–22, 58, 70, 73

Vale of White Horse 19–21, 59; District Council (VWHDC) 19–20, 70, 87, 90
voluntary 18–19
volunteer/s 43, 93, 97, 111, 113–114, 131–132

Whitecross 26–27, 41, 56, 74, 130
Wootton 22, 35, 60, 70
Wootton and St Helen Without Neighbourhood Plan (WSHWNP) 23–28, 73–82, 95–100, 129–132